NUMBERS:

RATIONAL AND IRRATIONAL

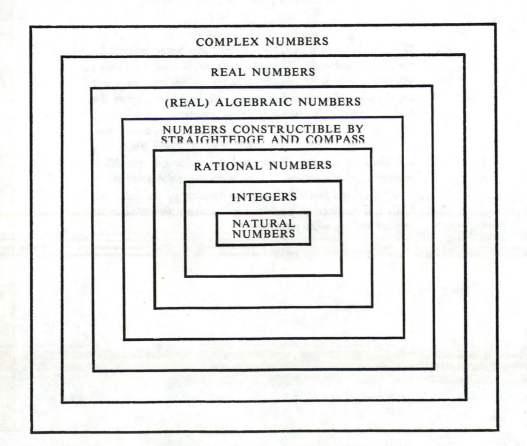

COMPLEX NUMBERS

REAL NUMBERS

(REAL) ALGEBRAIC NUMBERS

NUMBERS CONSTRUCTIBLE BY
STRAIGHTEDGE AND COMPASS

RATIONAL NUMBERS

INTEGERS

NATURAL
NUMBERS

NEW MATHEMATICAL LIBRARY

PUBLISHED BY

THE MATHEMATICAL ASSOCIATION OF AMERICA

Editorial Committee

Ivan Niven, Chairman (1976–77)   Anneli Lax, Editor
University of Oregon   *New York University*

W. G. Chinn (1977–79) *City College of San Francisco*
Basil Gordon (1977–79) *University of California, Los Angeles*
M. M. Schiffer (1976–78) *Stanford University*

The New Mathematical Library (NML) was begun in 1961 by the School Mathematics Study Group to make available to high school students short expository books on various topics not usually covered in the high school syllabus. In a decade the NML matured into a steadily growing series of some twenty titles of interest not only to the originally intended audience, but to college students and teachers at all levels. Previously published by Random House and L. W. Singer, the NML became a publication series of the Mathematical Association of America (MAA) in 1975. Under the auspices of the MAA the NML will continue to grow and will remain dedicated to its original and expanded purposes.

NUMBERS:

RATIONAL AND IRRATIONAL

by

Ivan Niven

University of Oregon

1

THE MATHEMATICAL ASSOCIATION
OF AMERICA

Fourteenth Printing

©Copyright, 1961, by the Mathematical Association of America
All rights reserved under International and Pan-American Copyright
Conventions. Published in Washington, D.C. by the
Mathematical Association of America

Library of Congress Catalog Card Number: 61—6226

Complete Set ISBN 0-88385-600-X

Vol. 1 0-88385-601-8

Manufactured in the United States of America

Note to the Reader

This book is one of a series written by professional mathematicians in order to make some important mathematical ideas interesting and understandable to a large audience of high school students and laymen. Most of the volumes in the *New Mathematical Library* cover topics not usually included in the high school curriculum; they vary in difficulty, and, even within a single book, some parts require a greater degree of concentration than others. Thus, while the reader needs little technical knowledge to understand most of these books, he will have to make an intellectual effort.

If the reader has so far encountered mathematics only in classroom work, he should keep in mind that a book on mathematics cannot be read quickly. Nor must he expect to understand all parts of the book on first reading. He should feel free to skip complicated parts and return to them later; often an argument will be clarified by a subsequent remark. On the other hand, sections containing thoroughly familiar material may be read very quickly.

The best way to learn mathematics is to *do* mathematics, and each book includes problems, some of which may require considerable thought. The reader is urged to acquire the habit of reading with paper and pencil in hand; in this way mathematics will become increasingly meaningful to him.

The authors and editorial committee are interested in reactions to the books in this series and hope that readers will write to: Anneli Lax, Editor, New Mathematical Library, NEW YORK UNIVERSITY, THE COURANT INSTITUTE OF MATHEMATICAL SCIENCES, 251 Mercer Street, New York, N. Y. 10012.

The Editors

NEW MATHEMATICAL LIBRARY

Other titles in preparation.

C O N T E N T S

NUMBERS:
RATIONAL AND IRRATIONAL

Preface to the Tenth Printing

In the Russian translation of this book published in the U.S.S.R. an additional appendix was added by the mathematician I. M. Yaglom, establishing the irrationality of almost all values of the trigonometric functions for rational degree arguments. It seemed to me that such an appendix should also be included in the English language edition, and I am grateful to Random House for agreeing at once to such an addition. Thus in this printing there is a new Appendix D, TRIGONOMETRIC NUMBERS, establishing the same results as in the Russian version, but with simpler methods I believe. A few minor changes have also been made, mostly clarifications in wording. I am indebted to several correspondents who suggested the improvements.

INTRODUCTION

The simplest numbers are the positive whole numbers, 1, 2, 3, and so on, used for counting. These are called *natural numbers* and have been with us for so many millennia that the famous mathematician Kronecker reputedly said: "God created the natural numbers; all the rest is the work of man."

The basic necessities of everyday life led to the introduction of common fractions like 1/2, 2/3, 5/4, etc.† Such numbers are called *rational numbers*, not because they are "reasonable," but because they are ratios of whole numbers.

We may think of the natural numbers as represented by dots along a straight line (Fig. 1), each dot separated by one unit of length from

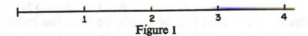

Figure 1

the previous one, as, for example, the number of inches along a tape measure. We may represent rational numbers along the same straight line (Fig. 2) and think of them as measuring fractions of length.

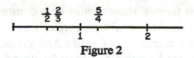

Figure 2

Much later, the Hindus invented the all-important number 0, and in the beginning of modern times Italian algebraists invented negative numbers. These may also be represented on a straight line, as shown in Fig. 3.

† For reasons of typography, the fractions $\frac{1}{2}$, $\frac{2}{3}$, $\frac{5}{4}$, and others in this book often appear with slanted bars, i.e. as 1/2, 2/3, 5/4.

3

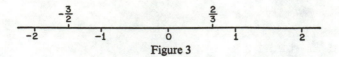

Figure 3

When mathematicians talk about rational numbers, they mean positive and negative whole numbers (which can be represented as ratios, e.g., 2 = 2/1 = 6/3, etc.), zero, and common fractions. The positive and negative whole numbers and zero are also called *integers*, therefore the class of rational numbers contains the class of integers.

The discovery that common fractions are not sufficient for the purposes of geometry was made by the Greeks more than 2500 years ago. They noticed, to their surprise and dismay, that the length of the

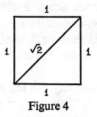

Figure 4

diagonal of a square whose sides are one unit long (Fig. 4) cannot be expressed by any rational number. (We shall prove this in Chapter 3.) Today we express this fact by saying that the square root of 2 (which, according to the Pythagorean Theorem, is the length of the diagonal of such a square) is an *irrational number*. What this means geometrically is that there is no common unit of length, no common mesh however fine, that can be put on both the side and the diagonal of a square a whole number of times. In other words, there is no unit of length, no matter how small, such that the side and the diagonal of a square are multiples of that unit. For the Greeks, this was an awkward discovery, because in many of their geometric proofs they had presumed that given any two line segments, there would be a common unit of length. Thus there was a gap in the logical structure of Euclidean geometry—an incompleteness in the discussion of ratios and proportions of lengths. In Section 3.7, we show how this gap may be closed and the theory of proportion made complete.

Similarly, the circumference of a circle is an irrational multiple, namely π, of the diameter. Other irrational numbers appear when we try to evaluate some of the basic functions in mathematics. For example, if we try to find the values of a trigonometric function, say sin x, when x has the value 60°, we are led to the irrational number $\sqrt{3}/2$; similarly, if we evaluate the logarithmic function log x, even for rational values of x, we usually are led to irrational numbers. Although the numbers listed in tables of logarithmic and trigonometric functions are ostensibly rational, actually they are only rational approximations of the true values, which are irrational with few exceptions. Clearly, then, irrational numbers occur in various natural ways in elementary mathematics.

The *real numbers* consist of all rational and irrational numbers, and form the central number system of mathematics. In geometry, any discussion of lengths, areas, or volumes leads at once to the real numbers. Geometry affords, in fact, a simple intuitive device for describing the real numbers, i.e., the numbers required to measure all possible lengths in terms of a given unit length. If again we consider the representation of numbers as points along a straight line, we find that, although any segment, no matter how small, contains infinitely many rational points, there are many other points (such as $\sqrt{2}$, π, etc.) which measure lengths that cannot be expressed by rational numbers. But once all real numbers are taken into account, every point on the line corresponds to exactly one real number and every real number corresponds to exactly one point on the line. The fact that *all* lengths can be expressed as real numbers is known as the *completeness property* of these numbers, and on this property depends the entire development of mathematical analysis.

Thus the real numbers are of two kinds, the rational and the irrational. There is another, much more recent separation of the real numbers into two categories, the *algebraic* numbers and the *transcendental* numbers. A real number is said to be algebraic if it satisfies some algebraic equation with integer coefficients. For example, $\sqrt{2}$ is an algebraic number because it satisfies $x^2 - 2 = 0$. If a number is not algebraic, it is said to be transcendental. From this definition, it is not clear that there are any transcendental, i.e., non-algebraic, numbers. In 1851, the French mathematician, Liouville, established that transcendental numbers exist. Liouville did this by exhibiting certain numbers which he proved to be non-algebraic. In Chapter 7 we shall follow Liouville's method to establish the existence of transcendentals.

Later in the 19th century, it was proved that π is a transcendental number, and this result settled an ancient geometric construction problem known as "squaring the circle." This is discussed in Chapter 5. Another advance in the 19th century was made by Cantor, a German mathematician, who established the existence of transcendental numbers by an entirely different approach. Although Cantor's method, in contrast to Liouville's, does not exhibit a transcendental number in explicit form, it has the advantage of demonstrating that, in a certain sense, many more numbers are transcendental than algebraic. Such a statement requires the comparison of infinite classes, since there are infinitely many algebraic numbers and infinitely many transcendental numbers. These ideas are somewhat removed from the main discussions in this book, so Cantor's proof of the existence of transcendental numbers is given in Appendix C.

The plan of the book is to present the natural numbers, integers, rational numbers, and real numbers in the first three chapters. Then in Chapter 4, a standard method is given for identifying irrational numbers. Chapter 5 deals with the so-called trigonometric and logarithmic numbers, that is, those numbers whose values are given approximately in tables of trigonometric and logarithmic functions. Chapter 6 treats the question of how closely it is possible to approximate irrational numbers by the use of rational numbers. This chapter is more difficult and more specialized than the earlier chapters. It is included to give some readers an opportunity to explore mathematical arguments of a new kind.

Chapter 7 and Appendix C offer two entirely independent proofs of the existence of transcendental numbers, Chapter 7 by the method of Liouville, Appendix C by the method of Cantor. The techniques are markedly different and the reader will be well rewarded if he follows each. The proof in Chapter 7 is laden with unavoidable technical details; and, even more than in the earlier chapters, the reader will have to use pencil and paper to follow the arguments. In fact, it is possible that the reader may find Chapters 1 through 5 not too troublesome, Chapter 6 rather difficult, and Chapter 7 virtually impossible. In such a case, it is suggested that the reader postpone the study of Chapter 7 until he has more mathematical experience. On the other hand, any reader who finds very little trouble in going through Chapters 1 to 5 might prefer to read Chapter 7 before Chapter 6. In fact, Chapter 7 is independent of the rest of the book except for one well-known result on inequalities given in Sect. 6.1.

Appendix C can be read independently of Chapter 7 except that the factor theorem, Theorem 7.2, is needed. If the reader is not familiar with the theory of sets, he will find the ideas of Appendix C very novel.

Appendix A, on the infinitude of prime numbers, is not essential to the arguments developed in this book; it is included because of its close relation to the main topic and because this elegant proposition dates back to Euclid. Appendix B on the fundamental theorem of arithmetic, on the other hand, is essential to our arguments, particularly to those of Chapters 4 and 5; the proof of this theorem has been relegated to an appendix because it is somewhat lengthy and difficult in comparison with the proofs in the first five chapters. The mathematically inexperienced reader can accept the fundamental theorem of arithmetic on faith.

There are many exercises at the ends of sections; the reader should try a good number of these to check his understanding of the text. (Mathematics cannot be learned by watching the other fellow do it!) Some of the problems are starred to indicate greater difficulty. The reader should not necessarily be unhappy if he cannot solve all of these. Often, his success depends on his mathematical maturity, that is, his acquaintance with a fairly broad collection of mathematical procedures from his other studies of mathematics. Answers to problems are given at the end of the book and also some suggestions for solving a few of the more difficult problems.

The system of real numbers, rational and irrational, can be approached at any one of several levels of *rigor*. (The word "rigor" is used technically in mathematics to denote the degree to which a topic is developed from a careful logical standpoint, as contrasted with a more *intuitive* position wherein assertions are accepted as correct because they appear somewhat reasonable or self-evident.) Our purpose is to present a first look at the subject, along fairly intuitive lines. Thus we offer no axioms or postulates as a basis for the study. The prospective mathematician into whose hands this book may find its way will one day want to examine a careful axiomatic development of the real number system. Why so? The reason is that our viewpoint here is so descriptive that it leaves some basic questions unanswered. For example, in Chapter 3 we say that the real numbers can be described in this way, that way, and the other way. But how can we be certain that these various ways are descriptions of the same system? To give a more concrete example of a question we do not answer in this book: How do we know that $\sqrt{2} \cdot \sqrt{3} = \sqrt{6}$ or that $\sqrt[3]{5} \cdot \sqrt[3]{7} = \sqrt[3]{35}$? To answer such questions, a precise definition of operations for irrational numbers must be given. This will not be done here since it is not so easy as it might appear, and it is best to postpone this type of treatment until the student not only has more mathematical skill but also has a greater appreciation of the nature and meaning of mathematical proof. As the American mathematician E. H. Moore said, "Sufficient unto the day is the rigor thereof."

"The nature and meaning of mathematical proof!" It is not possible here and now to give a precise description of what constitutes a proof, and herein lies one of the most puzzling bugbears for the beginning student of mathematics. If the nature of proof cannot be described or formulated in detail, how can anyone learn it? It is learned, to use an oversimplified analogy, in the same manner as a child learns to identify colors, namely, by observing someone else identify green things, blue things, etc., and then by imitating what he has observed. There may be failures at first caused by an inadequate understanding of the categories or patterns, but eventually the learner gets the knack. And so it is with the riddle of mathematical proof. Some of our discussions are intended to shed light on the patterns of proof techniques, and so to acquaint the reader with notions and methods of proof. Thus while we cannot give any sure-fire recipe for what is and what is not a valid proof, we do say some things about the matter, and hope that the reader, before he reaches the end of this book, will not only recognize valid proofs but will enjoy constructing some himself.

CHAPTER ONE

Natural Numbers and Integers

The number system of mathematics begins with the ordinary numbers used in counting,

$$1, 2, 3, 4, 5, 6, 7, 8, 9, 10, 11, 12, \ldots.$$

These are the positive whole numbers which are called the *natural numbers*. The smallest natural number is 1, but there is no largest natural number, because regardless of how large a number is chosen, there exist larger ones. Thus we say that there are infinitely many natural numbers.

If any two natural numbers are added, the result will be a natural number; for example, $4 + 4 = 8$ and $4 + 7 = 11$. Similarly, if any two are multiplied, the product will be a natural number; for example, $4 \times 7 = 28$. These two properties can be stated briefly by saying that the natural numbers are *closed under addition* and *closed under multiplication*. In other words, if we have a collection of objects (say the set of all natural numbers) and an operation (say addition) such that, no matter on which members of our set we operate (say 4 and 7), the result is again a member of the original collection, then we say that the set is closed under that operation. Suppose we consider only the numbers 1, 2, 3. This set of only three numbers is not closed under addition because $1 + 3 = 4$, and 4 is not a member of this set. When we speak of the set of natural numbers we shall mean the set of *all* natural numbers. If we wish to consider only some of them, we shall specify which ones we include in our set. Thus we have seen that the set of natural numbers is closed under addition but that the special set consisting of only the three natural numbers 1, 2, 3 is not.

The natural numbers are not closed under subtraction. In order to see this, we need only show that not every subtraction of one natural number from another yields a natural number. For example, if 7 is subtracted from 4 the result, -3, is not a natural number. Of course if 4 is subtracted from 7, the result is the natural number 3; according to our definition, however, we cannot say that a set of numbers is closed under subtraction unless the result of every possible subtraction is contained within that set. Similarly the natural numbers are not closed under division, because if 4 is divided by 7 for example, the result is the fraction 4/7 which is not a natural number.

It happens in many cases that two natural numbers can be divided to give a natural number as a result, for example 35 divided by 5 gives 7. In this case we say that 5 is an *exact divisor* of 35, or more briefly, that 5 is a *divisor* or *factor* of 35. Turning the statement around, we say that 35 is a *multiple* of 5. In general, let b and d denote any two natural numbers; if there is a third natural number q such that $b = dq$, then d is said to be a *divisor* of b, or b a *multiple* of d. In the example above we have $b = 35$ and $d = 5$, and of course q has the value 7. The letters d and q were chosen specifically, because they remind us of the words "divisor" and "quotient."

1.1 Primes

How many divisors does the number 35 have? The answer is four, as can be seen by listing all the divisors: 1, 5, 7, 35. The question was not difficult, because 35 is a relatively small natural number. But now consider the following question: how many divisors does the number 187 have? This is not so easy to answer, but when we try 1, 2, 3, etc., it turns out that again the answer is four, namely 1, 11, 17, 187. It might have taken a little effort for the reader to find the divisors 11 and 17, but the divisors 1 and 187 are obvious. Similarly it is apparent that 179 has divisors 1 and 179, and it turns out that these are the only divisors. When, as in the case of 179, a natural number has exactly two divisors, such a number is called a *prime* or a *prime number*. Another way of saying this is that *a prime is a natural number whose only divisors are itself and 1*. The first few primes in order of size are

$$2, 3, 5, 7, 11, 13, 17, 19, 23, 29, 31, 37, 41, 43, 47, \ldots .$$

Note that 1 is not listed as a prime. The fact that 1 is not a prime is a mathematical convention or agreement or, to say this another way, it is a matter of definition. Mathematicians have agreed not to call 1 a prime. The decision could have been made the other way, to include 1 among the primes. But with 1 excluded, it is possible to state propositions about primes without making exceptions or qualifications, as will be shown later.

Problem Set 1

[In the problem sets, the starred problems are the more difficult ones.]

1. Decide which of the following statements are true and which are false.
   (a) The set 1, 0, −1 is closed under addition.
   (b) The set 1, 0, −1 is closed under multiplication.
   (c) The set 1, 0, −1 is closed under subtraction.
   (d) The set of positive powers of 2, i.e., the set $2^1, 2^2, 2^3, 2^4, 2^5, 2^6, 2^7, \ldots$, is closed under multiplication
   *(e) The set of positive powers of 2 is closed under addition.

2. How many divisors does 30 have?

3. How many divisors does 16 have?

4. What is the smallest natural number having exactly three divisors?

5. Find all primes between 50 and 100.

*6. Prove that if 3 is a divisor of two numbers, it is a divisor of their sum and their difference. Generalize this and prove that if d is a divisor of two numbers b_1 and b_2, then d is a divisor of $b_1 + b_2$ and of $b_1 - b_2$.

1.2 Unique Factorization

The primes get scarcer as we consider larger and larger natural numbers. To illustrate what is meant by this, we point out that there are

168 primes between 1 and 1000,
135 primes between 1000 and 2000,
127 primes between 2000 and 3000,
120 primes between 3000 and 4000,
119 primes between 4000 and 5000.

Nevertheless, the list of primes is endless; that is, there are infinitely many prime numbers. This fact is proved in Appendix A at the end of the book. The proof does not require any special knowledge and so the reader can turn to it and read it now if he wishes. We have put this proof into an appendix because we do not need the result to establish any other proposition in this book. The proof is given because the result is interesting in itself.

Now, every natural number, except 1, either is a prime or can be factored into primes. For example, consider the natural number 94,860 which obviously is not a prime since

$$94{.}860 = 10 \times 9486.$$

Furthermore, 9486 is divisible by 2, also by 3, and, in fact, by 9. Thus we can write

$$94,860 = 10 \times 2 \times 9 \times 527$$
$$= 2 \times 2 \times 3 \times 3 \times 5 \times 527.$$

If 527 were a prime, the above expression would be a factoring of 94,860 into primes. But 527 is not a prime because $527 = 17 \times 31$. Hence we can write the prime factorization as

$$94,860 = 2 \times 2 \times 3 \times 3 \times 5 \times 17 \times 31.$$

We began with the particular number 94,860, but the procedure would also work no matter what natural number n we started with. For either n is a prime or it is not. If it is not, it can be factored into two smaller numbers, say a and b, so that $n = ab$. Each of the numbers a and b, in turn, either is a prime or can be factored into smaller numbers. Continuing this process, we finally factor n into primes completely.

The first sentence of the preceding paragraph distinguishes primes from other natural numbers. It is often desirable, in mathematics, to make definitions so general that a division into several cases becomes unnecessary. By "factorization into primes," for instance, we understood the representation of a number, say 12, as a product of *several* primes, in this case $2 \times 2 \times 3$. Now let us extend the meaning of "factorization into primes" so that it will include a single prime. For example, the prime number 23 then would have a prime factorization consisting of the single factor 23. With this extended meaning of "factoring into primes," our original statement can be replaced by the sentence: "Every natural number, except 1, can be factored into primes." Thus we have abbreviated the sentence and eliminated the necessity of distinguishing prime numbers from other numbers, at least for the purpose of making a statement about their factorization into primes.

It is a basic result in mathematics that the factoring of a natural number into primes *can be done in only one way*. For example, 94,860 cannot be factored into any primes other than the ones given above. Of course, the order of the factors can be different; for instance,

$$94,860 = 3 \times 17 \times 2 \times 5 \times 31 \times 3 \times 2.$$

But apart from such changes in the order, there is no other way of factoring 94,860. This result is known as the *Unique Factorization Theorem* or the *Fundamental Theorem of Arithmetic*, which is stated formally as follows:

THE FUNDAMENTAL THEOREM OF ARITHMETIC. *Every natural number, other than 1, can be factored into primes in only one way, apart from the order of the factors.*

This result is proved in Appendix B. It is a theorem that we shall use as we proceed with our discussion. The reason for putting the proof in an appendix is that it is somewhat complicated. However, no ideas occurring later in the book are used in the proof, so the reader may turn to Appendix B now if he wishes. Or he may postpone the study of Appendix B in order to take easier concepts first, harder concepts later.

The above statement of the Fundamental Theorem of Arithmetic provides one clue as to why 1 is not included among the prime numbers. For, if 1 were taken as a prime we then could write, for example,

$$35 = 5 \times 7 = 1 \times 5 \times 7,$$

and so 35 (or any other natural number) could be expressed in more than one way as a product of primes. Of course the Fundamental Theorem would still hold but its statement would require more qualifying phrases such as "except ..." or "unless ... ." Thus by banishing 1 from the list of primes, we can state our results more briefly and elegantly.

1.3 Integers

The natural numbers 1, 2, 3, 4, ... are closed under addition and multiplication, but not under subtraction or division. Closure under subtraction can be achieved in a set of numbers which is extended to include zero and the negatives:

$$0, -1, -2, -3, -4, \ldots.$$

These, taken together with the natural numbers, form the *integers* or whole numbers

$$\ldots, -5, -4, -3, -2, -1, 0, 1, 2, 3, 4, 5, \ldots.$$

The reader is probably familiar with the basic properties

$$a + b = b + a, \qquad ab = ba, \qquad a \cdot 0 = 0 \cdot a = 0,$$

$$(a + b) + c = a + (b + c), \quad (ab)c = a(bc), \qquad (-a)(-b) = ab,$$

$$a + 0 = 0 + a = a, \qquad a \cdot 1 = 1 \cdot a = a,$$

$$a(b + c) = ab + ac.$$

where a, b, c are any integers. These properties hold for all the number systems discussed in this book. It is not our intention to discuss the origins of these particular properties. Such a discussion would lead to a study of the foundations of the number system (which will be treated in one of the other books in this series) and away from the topic under consideration here. Our purpose is to derive various properties of numbers, especially irrational numbers, taking the foundations for granted.

The integers, then, are closed under addition, subtraction, and multiplication. They are not closed under division, because for example, the result of dividing 2 by 3 is not a whole number and hence leads us out of the class of integers.

Before we define the division of integers, let us examine the other operations and their results. When we consider addition of integers, we see not only that the sum of two integers is again an integer but also that there is only one integer which is this sum. For example, the sum of 3 and -1 is 2, not 5 and not anything else. We can express this fact by saying that, given two integers, there exists a *unique* third integer which is the sum of the other two. Similarly, for multiplication: given two integers, there is a unique third integer which is the product of the other two.

When we discussed the division of natural numbers, we saw that it was *not* always true that for any two given natural numbers, say b and d, there was a third natural number, their *quotient*, such that $b = dq$. However, whenever such a third natural number q does exist, it is clear that it is the only one, so we did not bother to say that q should be a unique natural number such that $b = dq$. When we define the same division concepts in the set of integers, however, we must include the requirement that the quotient be unique. We shall now analyze why this is necessary.

We must first agree that it is desirable to have only *one* answer to each of the questions: How much is $3 - 7$? How much is $(-2) \cdot (-3)$? How much is $8 \div 4$? In other words, we want to obtain a unique result for our operations. Next, let us see what happens when we consider division in the set of integers. Again, let b and d be given integers and define the quotient q to be an integer such that $b = dq$. For example, let $b = -12$ and $d = 3$. Clearly, $q = -4$ because $-12 = 3 \cdot (-4)$. An appropriate q exists and is unique. Next, let b be any integer and let d be the integer 0. We must find a q such that $b = 0 \cdot q$. If $b \neq 0$,† this equation cannot be solved; i.e., there is no q for which it is true. If $b = 0$, then the equation reads $0 = 0 \cdot q$ and is satisfied by any integer

† The symbol $\neq$ means "is not equal to."

q. In other words, if a solution q of $b = 0 \cdot q$ exists at all, it is *not* unique. Since unique results of arithmetic operations are important, we must construct a number system so that the quotient of two integers not only exists but also is unique. The scheme is simply not to allow division by zero. We now can state that an integer d is called a *divisor* of an integer b if there exists a unique integer q such that $b = dq$. (Then, by the above analysis $d \neq 0$.) Or we can say that a non-zero integer d is called a divisor of b if there exists an integer q such that $b = dq$. (Since we barred 0 as a possible divisor, the quotient will automatically be unique.)

In our earlier discussion, we asked the question: how many divisors does the number 35 have? At that time, the discussion was restricted to natural numbers and so the answer to this question was *four*, namely, 1, 5, 7, and 35. If now we interpret the question to mean that the divisors are to be integers, the answer is *eight*: ± 1, ± 5, ± 7, and ± 35.

Problem Set 2

1. Is -5 a divisor of 35?
2. Is 5 a divisor of -35?
3. Is -5 a divisor of -35?
4. Is 3 a divisor of -35?
5. Is 1 a divisor of -35?
6. Is 1 a divisor of 0?
7. Is 0 a divisor of 1?
8. Is 1 a divisor of 1?
9. Is 0 a divisor of 0?
10. Is 1 a divisor of every integer?
11. Is 0 a multiple of 35?
12. Verify that there are twenty-five primes between 1 and 100 and twenty-one primes between 100 and 200.

1.4 Even and Odd Integers

An integer is said to be *even* if it is divisible by 2; otherwise it is said to be *odd*. Thus the even integers are

$$\ldots, -8, -6, -4, -2, 0, 2, 4, 6, 8, \ldots$$

and the odd integers are

$$\ldots, -7, -5, -3, -1, 1, 3, 5, 7, \ldots.$$

Since an even integer is divisible by 2, we can write every even integer in the form $2n$, where the symbol n stands for any integer whatsoever. When a symbol (such as the letter n in our discussion) is permitted to represent any member of some specified set of objects (the set of integers in this case), we say that the *domain* of values of that symbol is the specified set. In the case under consideration, we say that every even integer can be written in the form $2n$, where the domain of n is the set of integers. For example, the even integers 18, 34, 12, and -62 are seen to have the form $2n$ when n is 9, 17, 6, and -31, respectively. There is no particular reason for using the letter n. Instead of saying that even integers are integers of the form $2n$, we could just as well say that they are integers of the form $2m$, or of the form $2j$, or of the form $2k$.

If two even integers are added, the result is an even integer. This is illustrated by the examples:

$$\begin{array}{cccc} 12 & 30 & 46 & -10 \\ \underline{14} & \underline{22} & \underline{-14} & \underline{-46} \\ 26 & 52 & 32 & -56 \end{array}$$

However, to *prove* the general principle that *the even integers are closed under addition* requires more than a collection of examples. To present such proof, we make use of the notation $2n$ for an even integer and $2m$, say, for another even integer. Then we can write the addition as

$$2m + 2n = 2(m + n).$$

The sum $2m + 2n$ has been written in the form $2(m + n)$ to exhibit its divisibility by 2. It would *not* have been sufficient to write

$$2n + 2n = 4n$$

because this represents the sum of an even integer and itself. In other words, we would have proved that twice an even integer is again even (in fact, divisible by 4) instead of proving that the sum of any two even integers is an even integer. Hence we used the notation $2n$ for one even integer and $2m$ for another to indicate that they are not necessarily the same.

What notation can we use to denote every odd integer? Note that whenever we add 1 to an even integer we get an odd integer. Thus we can say that every odd integer can be written in the form $2n + 1$. This form is not the only one. We could equally well have observed that whenever we subtract 1 from an even integer we get an odd integer. Thus we could say that every odd integer can be written in the form

$2n - 1$. For that matter, we can say that every odd integer can be written in the form $2n + 3$, in the form $2n - 3$, or in the form $2k - 5$, etc.

Can we say that every odd integer can be written in the form $2n^2 + 1$? If we substitute the integer values

$$\ldots, -5, -4, -3, -2, -1, 0, 1, 2, 3, 4, 5, \ldots$$

for n, we obtain the set of integers

$$\ldots, 51, 33, 19, 9, 3, 1, 3, 9, 19, 33, 51, \ldots$$

for $2n^2 + 1$. Each of these is odd, but they do not constitute *all* odd integers. For example, the odd integer 5 cannot be written in this form. Thus it is false that every odd integer can be written in the form $2n^2 + 1$, but it is true that any integer of the form $2n^2 + 1$ is odd. Similarly, it is false that every even integer can be written in the form $2k^2$, where the domain of k is the set of all integers; e.g., 6 is not equal to $2k^2$ no matter which integer is chosen for k. But it is true that any integer of the form $2k^2$ is even.

The relationship between these statements is the same as that between the statements "all cats are animals" and "all animals are cats." Clearly, the first is true and the second is not. This relationship will be discussed further when we examine statements which include the phrases "if," "only if," and "if and only if" (see Sect. 2.3).

Problem Set 3

Which of the following are true, and which are false? (It is understood that the domain of values for $n, m, j, \ldots$, is the set of all integers.)

1. Every odd integer can be expressed in the form

        (a) $2j - 1$.   (d) $2n^2 + 3$.
        (b) $2n + 7$.   (e) $2n^2 + 2n + 1$.
        (c) $4n + 1$.   (f) $2m - 9$.

2. Every integer of the form (a) above is odd; similarly for (b), (c), (d), (e), and (f).

3. Every even integer can be expressed in the form

        (a) $2n + 4$.   (d) $2 - 2m$.
        (b) $4n + 2$.   (e) $n^2 + 2$.
        (c) $2m - 2$.

4. Every integer of the form (a) in the previous problem is even; similarly for (b), (c), (d), and (e).

1.5 Closure Properties

The following two propositions will be of use in a succeeding chapter.

(*1*) *The set of even integers is closed under multiplication.*

(*2*) *The set of odd integers is closed under multiplication.*

To prove assertion (1), we must establish that the product of any two even integers is even. We can represent any two even integers symbolically by $2m$ and $2n$. Multiplying, we obtain

$$(2m)(2n) = 4mn = 2(2mn).$$

The product is divisible by 2, and so is even.

To prove assertion (2), we must establish that the product of two odd integers is odd. Representing the two odd integers by $2m + 1$ and $2n + 1$, we multiply these to get

$$(2m + 1)(2n + 1) = 4mn + 2m + 2n + 1 = 2(2mn + m + n) + 1.$$

Now $2(2mn + m + n)$ is even, whatever integers may be substituted for m and n in this expression. Hence $2(2mn + m + n) + 1$ is odd.

Assertions (1) and (2) could also be proved by an application of the unique factorization result, but we will not go into details about this alternative procedure. (The reader may find it challenging to try it by himself. Let him remember that an integer is even if and only if 2 occurs in its prime factorization.)

We have concentrated on even and odd integers, i.e., integers of the form $2m$ and of the form $2m + 1$. Evenness and oddness of integers are related to divisibility by 2. Analogous to this, we can consider the class of integers divisible by 3, namely

$$\ldots, -12, -9, -6, -3, 0, 3, 6, 9, 12, \ldots.$$

These are the multiples of 3. They can also be described as the class of integers of the form $3n$. The integers of the form $3n + 1$ are

$$\ldots, -11, -8, -5, -2, 1, 4, 7, 10, 13, \ldots,$$

and the integers of the form $3n + 2$ are

$$\ldots, -10, -7, -4, -1, 2, 5, 8, 11, 14, \ldots.$$

These three lists of integers include all integers; thus we can say that any integer is of exactly one of the forms $3n$, $3n + 1$, or $3n + 2$.

1.6 A Remark on the Nature of Proof

We said earlier that in order to prove that the even integers are closed under addition, i.e., that the sum of any two even integers is even, it would not suffice to examine only a few specific examples such as $12 + 14 = 26$. Since there are infinitely many even integers, we cannot check all cases of sums of specific pairs of even integers. So it is necessary to turn to some kind of algebraic symbolism; for example, the symbol $2n$, which can be used to express any even integer, enabled us to prove the closure of the set of all even integers under multiplication.

However, to prove a *negative* proposition such as "The odd integers are *not* closed under addition," we do not have to use any general algebraic symbols like $2m + 1$. The reason for this is that such a negative assertion can be established by a single example. To prove any statement which asserts that not all members of a set have a certain property, it clearly suffices to find one single member which does not have that property. To prove that not all boys have brown eyes, we need only find a blue-eyed or a hazel-eyed boy. To prove that not all sums of two odd integers are odd, observe that $3 + 5 = 8$, and this single case of the addition of two odd integers giving an even sum is sufficient proof. However, if we want to prove that the sum of *any* two odd integers is an even integer, it does not suffice to write $3 + 5 = 8$. Even if we write many cases, $7 + 11 = 18$, $5 + 53 = 58$, etc., we would not have a correct mathematical proof of the proposition.

Another example of a negative proposition is: "Not every prime number is odd." To prove this, we need merely point out that the even number 2 is a prime.

Problem Set 4

(The first three problems involve negative propositions and so can be solved by giving a single numerical example.)

1. Prove that the odd integers are not closed under subtraction.
2. Prove that the integers of the form $3n + 1$ are not closed under addition.
3. Prove that the integers of the form $3n + 2$ are not closed under multiplication.
4. Prove that the sum of any two odd integers is an even integer.
5. Prove that the following sets are closed under the indicated operation:
   (a) the integers of the form $3n + 1$, under multiplication;
   (b) the integers of the form $3n$, under addition;
   (c) the integers of the form $3n$, under multiplication.
6. Decide which of the following sets are closed under the indicated operation, and give a proof in each case:

(a) the integers of the form $6n + 3$, under addition;
(b) the integers of the form $6n + 3$, under multiplication;
(c) the integers of the form $6n$, under addition;
(d) the integers of the form $6n + 1$, under subtraction;
(e) the integers of the form $6n + 1$, under multiplication;
(f) the integers of the form $3n$, under multiplication;
(g) the integers not of the form $3n$, under multiplication.

CHAPTER TWO

Rational Numbers

2.1 Definition of Rational Numbers

We have seen that the natural numbers 1, 2, 3, 4, 5, ... are closed under addition and multiplication, and that the integers

$$\dots, -5, -4, -3, -2, -1, 0, 1, 2, 3, 4, 5, \dots$$

are closed under addition, multiplication, and subtraction. However, neither of these sets is closed under division, because division of integers can produce fractions like 4/3, 7/6, −2/5, etc. The entire collection of such fractions constitutes the rational numbers. *Thus a rational number (or a rational fraction) is a number which can be put in the form a/d, where a and d are integers, and d is not zero.* We have several remarks to make about this definition:

(1) We have required that d be different from zero. This requirement, expressed mathematically as $d \neq 0$, is necessary because d is in effect a divisor. Consider the instances:

Case (a) $a = 21,$ $d = 7,$ $\dfrac{a}{d} = \dfrac{21}{7} = \dfrac{3}{1} = 3;$

Case (b) $a = 25,$ $d = 7,$ $\dfrac{a}{d} = \dfrac{25}{7} = 3\dfrac{4}{7}.$

In case (a), d is a divisor in the sense of the preceding chapter; that is, 7 is an exact divisor of 21. In case (b), d is still a divisor, but in a different sense, because 7 is not an *exact* divisor of 25. But if we call 25 the *dividend*, and 7 the *divisor*, we get a *quotient* 3 and a *remainder* 4. Thus we are using the word *divisor*, in a more general sense, to cover a wider variety of cases than in Chapter 1. However, the divisor concept of Chapter 1 remains applicable in such instances as case (a) above; hence, as in Chapter 1, we must exclude $d = 0$.

(2) Note that while the terms *rational number* and *rational fraction* are synonymous, the word *fraction* alone is used to denote any algebraic expression with a numerator and a denominator, such as

$$\frac{\sqrt{3}}{2}, \quad \frac{17}{x}, \quad \text{or} \quad \frac{x^2 - y^2}{x^2 - y^2}.$$

(3) The definition of rational number included the words "a number which *can be put* in the form a/d, where a and d are integers, and $d \neq 0$." Why is it not enough to say "a number of the form a/d, where a and d are integers, $d \neq 0$"? The reason is that there are infinitely many ways to express a given fraction (for example, 2/3 also can be written as 4/6, 6/9, ... , or $2\pi/3\pi$, or $2\sqrt{3}/3\sqrt{3}$, or $-10/-15$, just to mention a few) and we do not want our definition of rational number to depend on the particular way in which somebody chooses to write it. A fraction is so defined that its value does not change if its numerator and denominator are both multiplied by the same quantity; but we cannot always tell, just by looking at a given fraction, whether or not it is rational. Consider, for example, the numbers

$$\frac{\sqrt{12}}{\sqrt{3}} \quad \text{and} \quad \frac{\sqrt{15}}{\sqrt{3}}$$

neither of which, as written, is in the form a/d, where a and d are integers. However, we can perform certain arithmetic manipulations on the first fraction and obtain

$$\frac{\sqrt{12}}{\sqrt{3}} = \frac{\sqrt{4 \cdot 3}}{\sqrt{3}} = \frac{2\sqrt{3}}{\sqrt{3}} = \frac{2}{1}.$$

We thus arrive at a number equal to the given fraction but of the specified form: $a = 2$, $d = 1$. Thus we see that $\sqrt{12}/\sqrt{3}$ is rational, but it would not have qualified had the definition stipulated that the number be in the right form to start with. In the case of $\sqrt{15}/\sqrt{3}$, the manipulations

$$\frac{\sqrt{15}}{\sqrt{3}} = \frac{\sqrt{5} \cdot \sqrt{3}}{\sqrt{3}} = \sqrt{5}$$

yield the number $\sqrt{5}$. We shall learn, in the following chapters, that $\sqrt{5}$ cannot be expressed as a ratio of integers and hence is irrational.

(4) Note that every integer is a rational number. We have just seen that this is so in the case of the integer 2. In general, the integers can be written in the form

$$\cdots, \quad \frac{-5}{1}, \quad \frac{-4}{1}, \quad \frac{-3}{1}, \quad \frac{-2}{1}, \quad \frac{-1}{1}, \quad \frac{0}{1}, \quad \frac{1}{1}, \quad \frac{2}{1}, \quad \frac{3}{1}, \quad \frac{4}{1}, \quad \frac{5}{1}, \quad \cdots,$$

where each is given the denominator 1.

Problem Set 5

1. Prove that the integer 2 can be written in rational form a/d (with integers a and d) in infinitely many ways.

2. Prove that the rational number $1/3$ can be written in rational form a/d in infinitely many ways.

3. Prove that the integer 0 can be written in rational form a/d in infinitely many ways.

4. Prove that every rational number has infinitely many representations in rational form.

5. *Definition.* Let k be any number; then the *reciprocal* of k is another number, say l, such that $k \cdot l = 1$.

   This definition has the consequence that all numbers except 0 have reciprocals. Given $k \neq 0$, by definition its reciprocal l satisfies the equation $k \cdot l = 1$; hence

   $$l = \frac{1}{k}$$

   which is meaningful only for $k \neq 0$. Prove that the reciprocal of any rational number (except zero) is rational.

2.2 Terminating and Non-terminating Decimals

There is another representation of the rational number $1/2$ which is different from the forms $2/4$, $3/6$, $4/8$, etc., namely as a decimal, 0.5. The decimal representations of some fractions are terminating, or finite; for example,

$$\frac{1}{2} = 0.5, \qquad \frac{2}{5} = 0.4, \qquad \frac{1}{80} = 0.0125.$$

Other fractions, however, have non-terminating, or infinite, decimal representations; for example,

$$\frac{1}{3} = 0.33333\cdots, \qquad \frac{1}{6} = 0.16666\cdots, \qquad \frac{5}{11} = 0.454545\cdots.$$

These infinite decimals can be obtained from the fractions by dividing the denominator into the numerator. In the case 5/11, for example, we divide 11 into $5.000\cdots$ and obtain the result $0.454545\cdots$.

Which rational fractions a/b have terminating decimal representations? Before we answer this question in general, let us examine an example, say the terminating decimal 0.8625. We know that

$$0.8625 = \frac{8625}{10000},$$

and that any terminating decimal can be written as a rational fraction with a denominator which is 10, 100, 1000, or some other power of 10. If the fraction on the right is reduced to lowest terms, we get

$$0.8625 = \frac{8625}{10000} = \frac{69}{80}.$$

The denominator 80 was obtained by dividing 10,000 by 125, the greatest common factor of 10,000 and 8625. Now the integer 80, like 10,000, has only the two prime factors 2 and 5 in its complete prime factorization. If we had started with any terminating decimal whatever, instead of 0.8625, the corresponding rational fraction form a/b in lowest terms† would have the same property. That is, the denominator b could have 2 and 5 as prime factors, but no others, because b is always a factor of some power of 10, and $10 = 2\cdot 5$. This turns out to be the deciding issue, and we shall prove the general statement:

A rational fraction a/b in lowest terms has a terminating decimal expansion if and only if the integer b has no prime factors other than 2 and 5.

It should be understood that b does not have to have 2 and 5 as prime factors; it may have only one or perhaps neither as prime factors:

$$\frac{1}{25} = 0.04, \qquad \frac{1}{16} = 0.0625, \qquad \frac{7}{1} = 7.0,$$

† A rational number a/b is in lowest terms if a and b have no common divisor d greater than 1.

where b has the values 25, 16, and 1. The important notion is that b must not have any prime factors *other than* 2 and 5.

Note that the above proposition contains the words *if and only if*. So far, all we have proved is the *only if* part because we showed that there would be a terminating expansion only if b is divisible by no primes other than 2 and 5. (In other words, if b is divisible by primes other than 2 and 5, then the fraction a/b in lowest terms will not have a terminating decimal expansion.)

The *if* part of the proposition states: if the integer b has no prime factors other than 2 and 5, then the rational fraction a/b in lowest terms has a terminating decimal expansion. To prove the *if* part, we must begin with any rational fraction a/b in lowest terms, assume that b has at most the prime factors 2 and 5, and prove that the corresponding decimal expression is of the terminating type. Let us first consider an example, say

$$\frac{a}{b} = \frac{9741}{3200} = \frac{9741}{2^7 \cdot 5^2}.$$

To convert this into a decimal, we merely change it into a fraction whose denominator is a power of 10. This can be achieved if we multiply both the numerator and denominator by 5^5:

$$\frac{9741}{2^7 \cdot 5^2} = \frac{9741 \cdot 5^5}{2^7 \cdot 5^7} = \frac{30440625}{10^7} = 3.0440625.$$

This argument can be generalized from this special case to any instance whatever in the following way: Suppose that b is of the form $2^m \cdot 5^n$, where m and n are positive integers or zero. Now, either n is less than or equal to m (written $n \leqq m$) or n is greater than m (written $n > m$). When $n \leqq m$, we multiply both the numerator and denominator of the fraction by 5^{m-n}:

$$\frac{a}{b} = \frac{a}{2^m \cdot 5^n} = \frac{a \cdot 5^{m-n}}{2^m \cdot 5^n \cdot 5^{m-n}} = \frac{a \cdot 5^{m-n}}{2^m \cdot 5^m} = \frac{a \cdot 5^{m-n}}{10^m}.$$

Since $m - n$ is positive or zero, 5^{m-n} is an integer, and so $a \cdot 5^{m-n}$ is also an integer, say c. Hence we can write

$$\frac{a}{b} = \frac{c}{10^m},$$

and since division of the integer c by 10^m merely requires that we insert the decimal point at the correct place, we get a terminating decimal.

On the other hand, if $n > m$, we would multiply the numerator and denominator of a/b by 2^{n-m}:

$$\frac{a}{b} = \frac{a}{2^m \cdot 5^n} = \frac{a \cdot 2^{n-m}}{2^m \cdot 5^n \cdot 2^{n-m}} = \frac{a \cdot 2^{n-m}}{2^n \cdot 5^n} = \frac{a \cdot 2^{n-m}}{10^n}.$$

Writing d for the integer $a \cdot 2^{n-m}$, we get

$$\frac{a}{b} = \frac{d}{10^n}.$$

Thus we have a terminating decimal as before.

Problem Set 6

1. Express the following fractions as terminating decimals:

(a) $\frac{1}{4}$, (b) $\frac{3}{200}$, (c) $\frac{321}{400}$, (d) $\frac{7}{625}$, (e) $\frac{352}{125}$, (f) $\frac{3149}{2500}$.

2.3 The Many Ways of Stating and Proving Propositions

We have used the phrase *if and only if* without precisely defining it. So at this point we pause in our discussion of rational numbers to explain briefly some of the language used in making mathematical statements, and also the relation of this language to the underlying logic. There are two basic kinds of assertions or propositions in mathematics:

> If A then B.
> If A then B, and conversely.

We take these up in turn.

When we state "if m and n are even integers, then mn is even," as we did in Section 1.5, we have an "if A then B" kind of assertion. Now, such an assertion may be stated in many ways, as illustrated by the following list:

Ways of Stating "If A then B"

[1] If A is true then B is true.
[2] If A holds then B holds.
[3] A implies B.
[4] B is implied by A.

[5] B follows from A.

[6] A is a sufficient condition for B.

[7] B is a necessary condition for A.

[8] B is true provided A is true.

[9] B is true if A is true.

[10] A is true only if B is true.

[11] It is impossible to have A true and B false at the same time.

[12] If B is false then A is false.

This list includes only the most common forms and is not complete, since there is virtually no limit to the number of possible forms of the statement. Some items, [6] and [7] for example, are not used in this book, but are included for completeness. All but [12] can be regarded as definitions of such terms as "implies," "necessary condition," "sufficient condition," and "only if."

Consider [10], for example, which defines the technical use in mathematics of the term "only if." Replacing the symbols A and B with the assertions about m and n of our previous discussion, we conclude that the following propositions both say the same thing.

"If the integers m and n are even, then the integer mn is even."

"The integers m and n are even only if the integer mn is even."

If the reader feels that they do not say the same thing, his feeling arises from some day-to-day usage of the word "only" to which he is accustomed. In this case, he should recognize a distinction between the technical language of mathematics and the everyday use of English. While these languages have much in common, there are pointed differences, as in the example under discussion. (After a person becomes skilled in the mathematical use of language, he can, if he chooses, use *it* as the form of his day-to-day speech. However, if he so chooses, he will be regarded by the man-in-the-street as pedantic, affected, or, at least, stuffy.)

What we have said thus far about the list of ways of saying "if A then B" is that forms [1] to [11] are based on nothing more than agreements about the way we use language in mathematics. Form [12] involves not only a new wording but also a fundamental axiom of logic. The fact that [12] says the same thing as "if A then B" has its basis in logic, and not simply in a different arrangement of words. The axiom of logic referred to (known as the *law of the excluded middle*) states that either A is true or A is false, where by A we mean any statement capable of analysis. In essence, the axiom excludes any middle ground between the truth and the falsity of A. Let us take this axiom for granted, and then prove that forms [1] and [12] say the same thing.

To do this, we must prove that [1] implies [12], and conversely, that [12] implies [1]. First, we assume [1] and then consider [12]:

"If B is false then A is false."

Is it possible that this conclusion is wrong, that it should be "A is true"? If this were so, then by use of [1] we would conclude that B is true, but this contradicts the hypothesis in [12]. Therefore, the conclusion "A is false" is right.

Conversely, let us assume [12] and prove that [1] follows:

"If A is true then B is true."

We ask whether this conclusion is wrong; should it be "B is false"? If this were so, then by use of [12] we would conclude that A is false, but this contradicts the hypothesis in [1]. Therefore "B is true" is the correct conclusion.

Forms [11] and [12] give the clue to the nature of indirect proof. Suppose we want to prove the assertion "if A then B." A direct proof is one in which the statement A is taken for granted, or assumed, and then the statement B is deduced. But if we examine form [11], we see that we can give a proof by assuming both the truth of A and the falsity of B, and then deducing a contradiction. This is a proof by contradiction, one of the methods of indirect proof. This type of proof can be spotted by noting the assumptions formulated in the proof; one is usually asked to assume first that the statement which is actually to be proved is false. Indirect proofs can also be spotted by the kind of language that occurs at the end of the proof, such as "... and so we have a contradiction and the theorem is proved."

Another common type of indirect proof is suggested by [12]. Thus to prove "if A then B," we can assume that B is false and then deduce that A is false. The three types of proof that we have identified are:

Assume A, deduce B. (direct proof)

Assume A true and B false, deduce a contradiction. (a form of indirect proof)

Assume B false, deduce that A is false. (another form of indirect proof)

Now, a curious thing about the way in which mathematics books are written (including the present one) is that these three types of proof are used freely, but often without any clear indication as to which type is being employed at any particular time! In effect, the reader is expected to solve a little riddle by identifying for his own thought processes the type of proof being used. The riddle is not a difficult one, however, and the reader usually can spot the assumptions made by the writer at the start of the proof.

Next, let us consider the second kind of mathematical proposition:

"If A then B, and conversely,"

which is mentioned at the beginning of this section. The words *and conversely* have the significance "if B then A," and this is the *converse* of "if A then B." The reader is probably aware that a statement and its converse are two different things. One may be true and the other false, both may be true, or both may be false, depending on the circumstances. For example, the statement "if m and n are even, then mn is even" is true, whereas the converse "if mn is even, then m and n are even" is false.

We now parallel our earlier list and indicate various ways of stating "if A then B, and conversely":

> If B then A, and conversely.
> A is true if and only if B is true.
> B is true if and only if A is true.
> A is false if and only if B is false.
> B is false if and only if A is false.
> A implies B, and conversely.
> B implies A, and conversely.
> A is a necessary and sufficient condition for B.
> B is a necessary and sufficient condition for A.
> A and B are equivalent statements.

All of these statements say the same thing.

But now let us note the wide variety of methods of proof available for establishing "if A then B, and conversely." As we saw earlier, there are three basic approaches to the proof of "if A then B." Similarly there are three possible methods of proof for "if B then A." Since any one of the first three may be combined with any one of the second three, there are nine possible organizations of the proof of "if A then B, and conversely." Perhaps the most common pattern is the direct proof each way: i.e.,

> (1) Assume A, deduce B.
> (2) Assume B, deduce A.

Another common pattern is this:

> (1) Assume A, deduce B.
> (2) Assume A false, deduce that B is false.

In somewhat complex proofs, these patterns are often combined. A proof of "if A then F" may be built by means of a chain of statements: "if A then B," "if B then C," "if C then D," "if D then E," "if E then F." In this case, each statement implies the next. Now, if each statement and its converse can be established by one of the above patterns, then we also have "if F then E," "if E then D," "if D then C," "if C then B," "if B then A," so that the converse "if F then A" of the original statement also holds. When an author says "the converse can be proved by reversing the steps," this is what he means.

All these patterns can be found in mathematical books, and as we said before, the writer will often launch forth on the proof of a theorem with no clear declaration of which pattern he is following. The writer expects the reader to figure out for himself the nature of the proof technique.

Problem Set 7

1. Prove that the statement "if mn is even, then m and n are even" is false.

2. Which of the following assertions are true, and which are false? The rational fraction a/b in lowest terms has a terminating decimal representation:
   (a) if and only if b is divisible by no prime other than 2;
   (b) if b is divisible by no prime other than 2;
   (c) only if b is divisible by no prime other than 2.
   (d) if and only if b is not divisible by 3;
   (e) if b is not divisible by 3;
   (f) only if b is not divisible by 3.

3. Which of the following statements are true, and which are false? The rational fraction a/b has a terminating decimal expansion:
   (a) if and only if b has no prime factors other than 2 and 5;
   (b) if b has no prime factors other than 2 and 5;
   (c) only if b has no prime factors other than 2 and 5.
   Suggestion. Observe that it has not been specified that a/b is in lowest terms.

4. A recent book on algebra† employs the following statement as an axiom: "$ab = 0$ only if $a = 0$ or $b = 0$." Rewrite this in the form "if A then B."

5. (a) Prove that if β (beta) is a rational number, then β^2 is also rational.
   (b) Does this amount to proving that if β^2 is irrational, then β is irrational?

2.4 Periodic Decimals

We now return to the topic of rational numbers. We have separated rational fractions into two types, i.e., those with terminating decimals

---

† W. W. Sawyer, *A Concrete Approach to Abstract Algebra*, p. 30.

and those with infinite decimals. We now can establish that each such infinite decimal has a repeating pattern such as

$$\frac{5}{11} = 0.454545\cdots \quad \text{and} \quad \frac{3097}{9900} = 0.31282828\cdots .$$

For convenience we will use the standard notation to indicate a periodic decimal, namely by the use of a bar over the repeating part:

$$\frac{5}{11} = 0.\overline{45}, \quad \frac{3097}{9900} = 0.31\overline{28}, \quad \frac{1}{3} = 0.\overline{3}, \quad \frac{1}{6} = 0.1\overline{6}, \quad \text{etc.}$$

The reason for the repetitive pattern of digits can be seen from a consideration of the standard method of converting a fraction, 2/7 for example, into decimal form:

$$
\begin{array}{r}
.285714 \\
7\overline{)2.000000} \\
1\,4 \\
\hline
60 \\
56 \\
\hline
40 \\
35 \\
\hline
50 \\
49 \\
\hline
10 \\
7 \\
\hline
30 \\
28 \\
\hline
2
\end{array}
$$

$$\frac{2}{7} = 0.\overline{285714}$$

In the division process, the successive remainders are 6, 4, 5, 1, 3, 2. When the remainder 2 is reached, the cycle is complete and we have a recurrence of the division of 7 into 20. The remainders are all less than the divisor 7, so there must be a recurrence since there are only six possible remainders. (The remainder 0 is ruled out of consideration because we are not examining terminating decimals.)

In the above example, the recurrence happened when the division of 7 into 20 turned up for a second time. Now, 7 into 20 was the first step in the whole division process. It need not happen that the first step is the one that recurs. Consider, for example, the conversion of 209/700 into decimal form:

$$
\begin{array}{r}
.29857142 \\
700\overline{)209.00000000} \\
140\ 0 \\
\hline
69\ 00 \\
63\ 00 \\
\hline
6\ 000 \\
5\ 600 \\
\hline
4000 \\
3500 \\
\hline
5000 \\
4900 \\
\hline
1000 \\
700 \\
\hline
3000 \\
2800 \\
\hline
2000 \\
1400 \\
\hline
600
\end{array}
$$

$$\frac{209}{700} = 0.29\overline{857142}$$

The recurrence here happens when we reach the remainder 600, which had occurred several steps earlier. With 700 as the divisor, we know that the possible remainders are the numbers 1, 2, 3, ..., 699. Thus we are sure of a repetition of a remainder although we might have to follow through quite a few steps to reach the repetition.

The general case, a/b, can be argued in a similar manner. For, when the integer b is divided into the integer a, the only possible remainders are 1, 2, 3, ..., $b - 2$, $b - 1$, and so a recurrence of the division process is certain. When the division process recurs, a cycle is started and the result is a periodic decimal.

What we have proved so far is half of the following proposition:

Any rational fraction a/b is expressible as a terminating decimal or an infinite periodic decimal; conversely, any decimal expansion which is either terminating or infinite periodic is equal to some rational number.

The converse deals with two kinds of decimals, terminating and infinite periodic. The terminating decimals have already been discussed and we have seen that they represent rational numbers. Let us now turn to the infinite periodic decimals. We first shall show, using a method which can be generalized to cover all cases, that a particular infinite periodic decimal is rational. After treating a particular case we shall apply the same method to any periodic decimal.

Consider the infinite periodic decimal

$$x = 28.123\overline{456} \quad \text{or} \quad x = 28.123456456 \cdots.$$

We shall multiply it first by one number and then by another; these numbers will be so chosen that, when we take the difference of the two products, the infinite periodic part will have been subtracted out. In the example, the numbers 10^6 and 10^3 will serve this purpose because

$$10^6 \cdot x = 28123456.\overline{456}$$

and

$$10^3 \cdot x = \quad\quad 28123.\overline{456},$$

so that the difference $10^6 \cdot x - 10^3 \cdot x$ is

$$999000x = 28095333.$$

Therefore

$$x = \frac{28095333}{999000},$$

which exhibits the fact that x is a rational number.

In generalizing this method we shall show that the numbers 10^3 and 10^6 were not "pulled out of a hat" but were chosen systematically. We shall omit the integer part of the decimal (i.e., the part that corresponds to 28 in the above example) because it plays no decisive role in the process. Thus we may write any repeating decimal (without integer part) in the form†

$$x = .a_1 a_2 \cdots a_s \overline{b_1 b_2 \cdots b_t},$$

where a_1, a_2, ..., a_s represent the s consecutive digits in the non-repeating part and b_1, b_2, ..., b_t represent the t digits in the repeating

---

† Note that the notation $a_1 a_2 \cdots a_s b_1 b_2 \cdots b_t$ used here is *not* the usual algebraic notation and does *not* denote the product of the numbers a_1, a_2, $\cdots$, b_t; in this proof it means the integer whose digits are a_1, a_2, $\cdots$, b_t. Furthermore, the symbols $1, 2, \cdots, s$ in the notation a_1, a_2, $\cdots$, a_s are called "subscripts" and have no significance except as identification tags; without subscripts we would soon run out of letters.

part. (In the above example, $s = 3$, $t = 3$; $a_1 = 1$, $a_2 = 2$, $a_3 = 3$, $b_1 = 4$, $b_2 = 5$, and $b_3 = 6$.) If we multiply x first by 10^{s+t}, then by 10^s, and subtract, we obtain

$$10^{s+t} \cdot x = a_1a_2 \cdots a_sb_1b_2 \cdots b_t + .\overline{b_1b_2 \cdots b_t},$$
$$10^s \cdot x = a_1a_2 \cdots a_s + .\overline{b_1b_2 \cdots b_t};$$

and

$$(10^{s+t} - 10^s) \cdot x = a_1a_2 \cdots a_sb_1b_2 \cdots b_t - a_1a_2 \cdots a_s,$$

so that

$$x = \frac{a_1a_2 \cdots a_sb_1b_2 \cdots b_t - a_1a_2 \cdots a_s}{10^{s+t} - 10^s},$$

which is of the integer-divided-by-integer form. Hence it is rational as we set out to prove.

Problem Set 8

1. Find rational numbers equal to the following decimals:

    (a) $0.111 \cdots$ (b) $5.6666 \cdots$ (c) $0.37\overline{43}$

    (d) $0.9\overline{987}$ (e) $0.000\overline{1}$ (f) $0.\overline{9}$

2.5 Every Terminating Decimal Can Be Written As a Periodic Decimal

What we have established in this chapter is that some rational numbers can be expressed as terminating decimals, whereas other rational numbers turn out to be infinite or non-terminating decimals. Curiously enough, every terminating decimal (except zero) can be expressed in a non-terminating form. Of course this can be done in a very obvious way when we write 6.8 as $6.8000 \cdots$, i.e., with an infinite succession of zeros. But apart from this obvious process of changing a terminating decimal into a non-terminating one by appending a whole string of zeros, there is another way that is a little surprising. Let us begin with the well-known decimal expansion for 1/3:

$$\frac{1}{3} = 0.33333 \cdots.$$

If we multiply both sides of this equation by 3, we get the strange-looking result

(1) $1 = 0.99999 \cdots.$

Thus we have equality between the terminating decimal 1, or 1.0, and the nonterminating decimal $0.99999\cdots$.

Let us look at equation (1) in another way. Suppose we denote the infinite decimal $0.99999\cdots$ by x; that is,

(2) $$x = 0.99999\cdots.$$

Multiplying by 10, we get

$$10x = 9.99999\cdots = 9 + 0.99999\cdots.$$

Subtracting eq. (2) from this, we obtain

$$9x = 9 \quad \text{or} \quad x = 1.$$

Thus we have proved eq. (1) by a different approach from that used initially.

Now, upon division by 10, 100, 1000, 10,000 etc., eq. (1) yields the entire succession of results

(3)
$$0.1 = 0.099999\cdots$$
$$0.01 = 0.0099999\cdots$$
$$0.001 = 0.00099999\cdots$$
$$0.0001 = 0.00009999\cdots, \text{ etc.}$$

These results can be used to convert any terminating decimal into a non-terminating form. For example, we can write

$$6.8 = 6.7 + 0.1 = 6.7 + 0.099999\cdots = 6.799999\cdots.$$

Some further examples are:

$$0.43 = 0.42 + 0.01 = 0.42 + 0.0099999\cdots = 0.4299999\cdots;$$

$$0.758 = 0.757 + 0.001 = 0.757 + 0.00099999\cdots = 0.75799999\cdots;$$

$$0.102 = 0.101 + 0.001 = 0.101 + 0.00099999\cdots = 0.10199999\cdots;$$

$$6.81 = 6.8 + 0.01 = 6.8 + 0.0099999\cdots = 6.8099999\cdots.$$

This device enables us to write any terminating decimal in non-terminating form. Conversely, eqs. (1) and (3) can be used to convert any decimal that has an infinite succession of nines into a terminating decimal:

$$0.4699999 \cdots = 0.46 + 0.0099999 \cdots = 0.46 + 0.01 = 0.47,$$

$$18.099999 \cdots = 18. + 0.099999 \cdots = 18. + 0.1 = 18.1.$$

The question of how many representations of a given number there are by decimals involves a matter of interpretation. For, in addition to writing 0.43 as 0.42999 $\cdots$, we can also write this number in the forms

$$0.430, \quad 0.4300, \quad 0.43000, \quad 0.430000, \quad \cdots.$$

These, however, are such trivial variations on 0.43 itself that we do not count them as essentially different representations. When we refer to the infinite decimal form of a number such as 0.43, we mean 0.42999 $\cdots$ and not 0.43000 $\cdots$.

Problem Set 9

1. Write each of the following as a terminating decimal:
   (a) 0.11999 $\cdots$ (b) 0.299999 $\cdots$ (c) 4.79999 $\cdots$ (d) 9.999 $\cdots$
2. Write each of the following as a non-terminating decimal:
   (a) 0.73   (b) 0.0099   (c) 13
3. Which rational numbers a/b have two essentially different decimal representations?
4. Which rational numbers a/b have three essentially different decimal representations?

2.6 A Summary

We have distinguished two types of rational numbers a/b, those in which the integer b has no factors other than 2 and 5, and all others. (It is presumed that a/b is in lowest terms.) Those of the first type can be written both as finite and infinite decimals; for example,

$$\frac{1}{2} = 0.5 = 0.499999 \cdots.$$

The numbers of the second type can be written only in infinite decimal form; for example,

$$\frac{1}{3} = 0.33333 \cdots.$$

These representations are the only possible ones in the sense that 1/2 and 1/3 cannot be expressed in any other decimal form excluding, of course, such trivialities as 0.500. We shall explain in the next chapter why this is so.

The emphasis has been on rational numbers and their decimal representations. Turning the matter around, let us reflect on decimal representations for a moment. All the infinite decimal expansions in this chapter have been periodic. What about non-periodic decimals such as

$$q = 0.101\,001\,000\,100\,001\,000\,001\,000\,000\,1 \cdots\cdots$$

formed by a series of ones separated by zeros, first one zero, then two zeros, then three zeros, and so on? What kind of a number, if it is a number, is q? From our studies in the present chapter we know that q is not a rational number. In the next chapter we shall broaden our inquiry to include such numbers as q.

CHAPTER THREE

Real Numbers

3.1 The Geometric Viewpoint

When coordinates are introduced in geometry, one straight line is designated as the *x*-axis, say, and this axis is graduated so that each point is associated with a number. This is done by taking two arbitrary (but distinct) points on the line as the positions for 0 and 1 such that the distance between these two points becomes the *unit of length,* or the *unit length.* It is conventional (Fig. 5) to take the 1-point to the

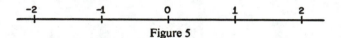

Figure 5

right of the 0-point, so that the points to the left of the 0-point are associated with negative numbers. The 0-point is called the *origin.* The point belonging to the number 7, for example, is seven units of length to the right of the origin. The point belonging to −7 is seven units to the left of the origin. In this way a number is attached to each point, the number being the distance from the point to the origin either with a plus sign if the point is to the right of the origin or with a minus sign if it is to the left. As shown in Fig. 6, rational numbers like −4/3, 1/2, and 2.3 are readily located by their relation to the unit length.

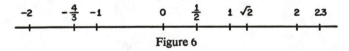

Figure 6

The symbol $\sqrt{2}$ denotes a number which, when multiplied by itself, yields 2; that is, $\sqrt{2} \cdot \sqrt{2} = 2$. To see the geometric meaning of $\sqrt{2}$, we consider a unit square as shown in Fig. 7, and we find from the Pythagorean Theorem that the square of the length of its diagonal is 2.

38

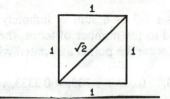

Figure 7. A square with sides of length 1

Hence, we denote the length of the diagonal by $\sqrt{2}$ and associate the number $\sqrt{2}$ with that point on the line whose distance from the origin is equal to the length of the diagonal of our unit square.

Since each point on the axis lies at some distance from the origin, it is intuitively clear that there is a number associated with each such point. By the *real numbers* we mean the collection of all the numbers associated with all the points. Every rational number is included because there is an appropriate distance from the origin for each rational number. Thus we can say that the rational numbers form a subclass of the real numbers.

However, there are real numbers which are not rational. The number $\sqrt{2}$ is not rational, as we shall prove later in this chapter. Any real number, such as $\sqrt{2}$, which is not rational is said to be *irrational*. Because of the way the definitions have been made, any real number is either rational or irrational. The straight line, or axis, with a number attached to each point in the manner described above, is called the *real line*. The points on this line are referred to as rational or irrational points depending on whether the corresponding numbers are rational or irrational.

Note that the above definition of an irrational number amounts to this: any real number which cannot be expressed as the ratio a/b of two integers is called an irrational number.

3.2 Decimal Representations

The number 1/3 is easily located on the real line at a point of trisection between the zero- and unit-points (Fig. 8). Now consider the

Figure 8

decimal representation of 1/3:

$$\frac{1}{3} = 0.33333\cdots = \frac{3}{10} + \frac{3}{100} + \frac{3}{1000} + \cdots.$$

This equation expresses 1/3 as a sum of infinitely many terms. Even though there is no end to the number of terms, the sum has a definite value, i.e., 1/3. If we locate the points associated with

$$0.3, \quad 0.33, \quad 0.333, \quad 0.3333, \ldots$$

on the real line, we get a sequence of points which converge on the point 1/3. This is illustrated in Fig. 9, where the unit of length has been

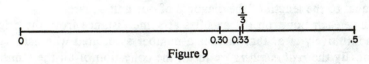

Figure 9

magnified. In the same way, any infinite decimal belongs to a particular point on the real line. In the case of the infinite decimal $0.99999 \cdots$, the point it represents is converged on by the points associated with

$$0.9, \quad 0.99, \quad 0.999, \quad 0.9999, \quad 0.99999, \text{ etc.}$$

As shown in Fig. 10, these points are converging on the point 1, in

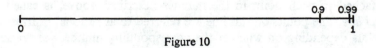

Figure 10

accordance with the equation $1 = 0.99999 \cdots$ of the preceding chapter.

Now, when we turn to the number

$$q = 0.101\,001\,000\,100\,001\,000\,001\,000\,000\,1 \cdot \cdot \cdots,$$

which was used as a previous example, we find that this number also belongs to a particular point on the real line. This point can be thought of as being converged on by the following chain of points:

0.1,

0.101,

0.101 001,

0.101 001 0001,

0.101 001 000 100 001, etc.

Since q is a non-periodic decimal, it is an irrational number, and the corresponding point is an irrational point.

This suggests another way of interpreting real numbers. The real numbers are the collection of all decimal expansions, finite or infinite, such as

$$17.34, \qquad 2.176, \qquad -6.037\,222\,22\cdots, \qquad q = \cdot 101\,001\,0001\cdots.$$

According to our studies in the preceding chapter, we can separate these decimal expansions into rational and irrational numbers. The rational numbers are those decimals which are either terminating or periodic; the irrational numbers are those which are non-periodic, such as the number q above. Moreover, since we have seen that every terminating decimal (or every decimal like $0.43000\cdots$ with an infinite succession of zeros) can also be written as a genuinely infinite periodic decimal, we can agree to write, within this section, all rational numbers as infinite periodic decimals. (By such an agreement we would, for example, write 0.43 in the form $0.42999\cdots$; this may seem awkward, but it will simplify the discussion below.)

We shall now show that *real numbers have a unique representation as infinite decimals*. That is to say, *two infinite decimals are equal only if they are identical, digit by digit*.

Why is the infinite decimal representation unique? We answer this question as follows: consider two numbers with different infinite decimal representations. Since the representations are different, there is at least one digit wherein this difference can be actually observed; for example,

$$a = 17.923416\cdots,$$

$$b = 17.923415\cdots.$$

The infinite succession of digits that follow 6 in the number a may be any collection that the reader chooses to imagine, except an infinite succession of zeros. A similar remark applies to the number b. Now, the fact that an infinite succession of zeros is excluded tells us that a is definitely larger than 17.923416, which in symbols is expressed as

$$a > 17.923416.$$

On the other hand, b is at most 17.923416, because we can have $b = 17.923416$ only when the succession of digits in b following the

"5" are all nines, i.e., when $b = 17.9234159$. The statement that b is at most 17.923416 is written symbolically as

$$b \leq 17.923416 \qquad \text{or} \qquad 17.923416 \geq b.$$

These inequalities for a and b state that

$$a > 17.923416 \geq b$$

and hence $a > b$. We have concluded, then, that a is greater than b, and, of course, this rules out the possibility of equality. Our argument has been applied to a special case of two particular numbers a and b but the reasoning generalizes at once to any pair of numbers having different infinite decimal representations.

3.3 The Irrationality of $\sqrt{2}$

We now give the traditional indirect proof that $\sqrt{2}$ is irrational, and in the next chapter we shall give yet another proof by means of a much more general approach.

In Chapter 1 we showed that the even integers are closed under multiplication, and likewise the odd integers. In particular, the square of an even integer is even and the square of an odd integer is odd.

Now suppose that $\sqrt{2}$ were a rational number, say

$$\sqrt{2} = \frac{a}{b},$$

where a and b are integers. We will presume, and this is essential for the argument, that the rational fraction a/b is in its lowest terms. Specifically, we shall make use of the fact that a and b are not both even, because if they were the fraction would not be in lowest terms. Squaring the above equation, and simplifying, we get

$$2 = \frac{a^2}{b^2}, \qquad a^2 = 2b^2.$$

The term $2b^2$ represents an even integer, so a^2 is an even integer, and hence a is an even integer, say $a = 2c$, where c is also an integer. Replacing a by $2c$ in the equation $a^2 = 2b^2$, we obtain

$$(2c)^2 = 2b^2, \qquad 4c^2 = 2b^2, \qquad 2c^2 = b^2.$$

The term $2c^2$ represents an even integer, so b^2 is an even integer, and hence b is an even integer. But now we have concluded that both a and b are even integers, whereas a/b was presumed to be in lowest terms. This contradiction leads us to conclude that it is not possible to express $\sqrt{2}$ in the rational form a/b, and therefore $\sqrt{2}$ is irrational.

3.4 The Irrationality of $\sqrt{3}$

One of the proofs that $\sqrt{3}$ is irrational is similar to the proof of the irrationality of $\sqrt{2}$ just given, except that here the key is divisibility by 3 rather than by 2. As a preliminary to the proof we establish that *the square of an integer is divisible by 3 if and only if the integer itself is divisible by 3*. To see this, we note that an integer divisible by 3 has the form $3n$, whereas an integer not divisible by 3 is of the form $3n + 1$ or the form $3n + 2$. Then the equations

$$(3n)^2 = 9n^2 = 3(3n^2),$$

$$(3n + 1)^2 = 9n^2 + 6n + 1 = 3(3n^2 + 2n) + 1,$$

$$(3n + 2)^2 = 9n^2 + 12n + 4 = 3(3n^2 + 4n + 1) + 1$$

confirm the above assertion.

Next suppose that $\sqrt{3}$ were a rational number, say

$$\sqrt{3} = \frac{a}{b},$$

where a and b are integers. Again, as in the $\sqrt{2}$ case, we presume that a/b is in lowest terms, so that not both a and b are divisible by 3. Squaring and simplifying the equation, we obtain

$$3 = \frac{a^2}{b^2}, \qquad a^2 = 3b^2.$$

The integer $3b^2$ is divisible by 3; that is, a^2 is divisible by 3. So a itself is divisible by 3, say $a = 3c$, where c is an integer. Replacing a by $3c$ in the equation $a^2 = 3b^2$, we get

$$(3c)^2 = 3b^2, \qquad 9c^2 = 3b^2, \qquad 3c^2 = b^2.$$

This shows that b^2 is divisible by 3, and hence b is divisible by 3. But we have established that both a and b are divisible by 3, and this is contrary to the presumption that a/b is in lowest terms. Therefore $\sqrt{3}$ is irrational.

3.5 Irrationality of $\sqrt{6}$ and $\sqrt{2} + \sqrt{3}$

The proofs of the irrationality of $\sqrt{2}$ and $\sqrt{3}$ depended on divisibility properties of integers by 2 and by 3, respectively, but the corresponding proof for $\sqrt{6}$ can be made to depend on divisibility either by 2 or by 3. For example, if we parallel the $\sqrt{2}$ proof, we would assume that

$$\sqrt{6} = \frac{a}{b},$$

where the integers a and b are not both even. Squaring, we would obtain

$$6 = \frac{a^2}{b^2}, \quad a^2 = 6b^2.$$

Now, $6b^2$ is even, so a^2 is even, so a is even, say $a = 2c$. Then we can write

$$a^2 = 6b^2, \quad (2c)^2 = 6b^2, \quad 4c^2 = 6b^2, \quad 2c^2 = 3b^2.$$

This tells us that $3b^2$ is even, so b^2 is even, and thus b is even. But a and b were presumed to be not both even, and so $\sqrt{6}$ is irrational. The reader may, as an exercise, deduce the same conclusion by means of a proof which is analogous to the $\sqrt{3}$ proof.

As a last example of irrationality in this chapter, we treat the case $\sqrt{2} + \sqrt{3}$ by making it depend on the $\sqrt{6}$ case. Suppose that $\sqrt{2} + \sqrt{3}$ were rational, say r, so that

$$\sqrt{2} + \sqrt{3} = r.$$

Squaring and simplifying, we get

$$2 + 2\sqrt{6} + 3 = r^2, \quad 2\sqrt{6} = r^2 - 5, \quad \sqrt{6} = \frac{r^2 - 5}{2}.$$

Now, rational numbers are closed under the four operations of addition, subtraction, multiplication, and division (except by zero), and so $\frac{1}{2}(r^2 - 5)$ is a rational number. But $\sqrt{6}$ is irrational, and thus we have a contradiction. We can therefore conclude that $\sqrt{2} + \sqrt{3}$ is irrational.

Given any integer $n = a \cdot b$ of which we know that $\sqrt{n} = \sqrt{a \cdot b}$ is irrational, we can deduce that the expression $\sqrt{a} + \sqrt{b}$ is irrational by imitating the above proof.

Problem Set 10

1. Give two proofs that the square of an integer is divisible by 5 if and only if the integer itself is divisible by 5.
   (a) First, give a proof parallel to the analysis in the text in the case of divisibility by 3. Start from the fact that every integer is of one of the five forms, $5n$, $5n + 1$, $5n + 2$, $5n + 3$, or $5n + 4$.
   (b) Next, give a proof by use of the Fundamental Theorem of Arithmetic. This theorem can be found in Chapter 1 or in Appendix B.

2. Prove that $\sqrt{5}$ is irrational.

3. Prove that $\sqrt{15}$ is irrational.

4. Prove that $\sqrt{5} + \sqrt{3}$ is irrational.

5. Prove that $\sqrt[3]{2}$ is irrational.

6. Given that α (alpha) is an irrational number, prove that $\alpha^{-1} = 1/\alpha$ also is irrational.

7. Is 0 rational or irrational?

3.6 The Words We Use

The language that we use to describe the various classes of numbers is part of our historical inheritance, and so it is not likely to change even though we may feel that some of the words are slightly peculiar. For example, in everyday speech when we describe something as "irrational," we usually mean that it is detached from good sense and therefore unreasonable. But of course we do not regard irrational numbers as being unreasonable. Apparently, the Greeks were surprised when they discovered irrational numbers, because they had felt that, given any two straight-line segments such as the side and the diagonal of a square, there would be some integers a and b so that the ratio of the lengths of the segments would be a/b. Thus the word "rational" in its mathematical sense has reference to this ratio of whole numbers and "irrational" refers to the absence of any such ratio.

The word "commensurable" also has been used to describe two lengths whose ratio is a rational number. Two *commensurable* lengths are so related that one can be "measured" by means of the other in the following sense: If there exists some integer k such that when the first segment is divided into k equal parts, each part of length l, it turns

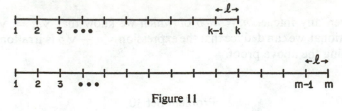

Figure 11

out that the second segment is measured by a whole number, say m, of such parts of length l, then the ratio of the lengths of the two segments is

$$\frac{kl}{ml} = \frac{k}{m},$$

i.e., rational (see Fig. 11). However, if the segments are such that the ratio of their lengths is irrational (e.g. the side and the diagonal of a square), then the above construction can never be performed, no matter how large we take k (and how small we take l)! In this case the given segments are said to be *incommensurable*.

Numbers like $\sqrt{2}$, $\sqrt[3]{24}$, or, in general, of the form $\sqrt[n]{a}$, where a is rational and n is an integer, are called *radicals*.

The term "real numbers" is another inheritance from the past. If we were to name them today, we would perhaps call them "one-dimensional numbers." In any event, we do not regard numbers beyond the range of real numbers as being "unreal." The reader is probably familiar with the complex numbers, of which the real numbers constitute a subclass. A complex number is a number of the form $a + bi$, where a and b are real and i satisfies the quadratic formula $i^2 = -1$. This definition is introduced merely to round out the discussion of classes of numbers. The scope of this book is limited to real numbers and so we are not concerned with the wider class of complex numbers.

3.7 An Application to Geometry

Most high school textbooks on geometry leave a gap in some proof or proofs where irrational numbers arise. The gap occurs when a result is proved in the rational case only, with the irrational case left unfinished. This happens often with the following result.

THEOREM 3.1. *If three parallel lines are cut by two transversals, with intersection points A, B, C, A', B', C', as shown in Fig. 12, then*

$$\frac{AB}{BC} = \frac{A'B'}{B'C'},$$

where, for example, AB denotes the length of the line segment from A to B.

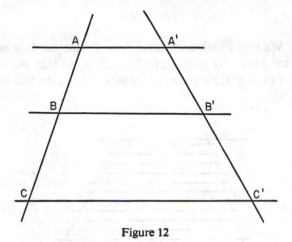

Figure 12

This theorem can be used to prove the basic theorem on similar triangles: *if the three angles of one triangle are equal, respectively, to the three angles of another, then corresponding sides are proportional*

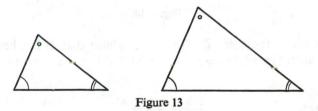

Figure 13

(Fig. 13). This result, in turn, is often used to prove Pythagoras' Theorem, and so trigonometry and analytic geometry are built on the basis of these theorems.

We shall now prove Theorem 3.1 for the case where AB/BC is irrational. We shall take Theorem 3.1 for granted in the case where AB/BC is rational, because this part of the theorem ordinarily is proved in books on elementary geometry. Before proving Theorem 3.1 in the case where AB/BC is irrational, it will be helpful to establish the following preliminary result.

THEOREM 3.2. *If m and n are positive integers such that*

$$\frac{m}{n} < \frac{AB}{BC},$$

then

$$\frac{m}{n} < \frac{A'B'}{B'C'}.$$

PROOF. We begin with a construction. Divide the line segment BC into n equal parts, say each part of length α, so that $BC = n\alpha$. Then mark off m more of these pieces of lengths α along the line segment BA,

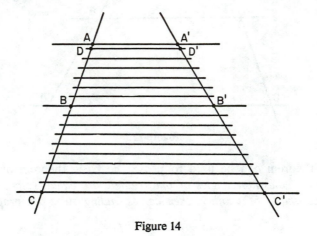

Figure 14

terminating at the point D. We first establish that D lies between B and A, as in Fig. 14. Since $BC = n\alpha$ and $DB = m\alpha$, we can write

$$\frac{DB}{BC} = \frac{m\alpha}{n\alpha} = \frac{m}{n};$$

by assumption

$$\frac{m}{n} < \frac{AB}{BC},$$

and so

$$\frac{DB}{BC} < \frac{AB}{BC}.$$

This last inequality implies that $DB < AB$, because both fractions have the same denominator BC. Thus, since DB is shorter than AB, it follows that D lies inside the line segment AB.

Next, from all these points of division we draw lines parallel to AA', with the point D' corresponding to D on the right-hand side as in

Fig. 14. Thus by Theorem 3.1 for the rational case (which we took for granted), $B'C'$ is divided into n equal parts, and $D'B'$ into m equal parts of the same length, so that

$$\frac{D'B'}{B'C'} = \frac{m}{n}.$$

However, from Fig. 14 we note that $D'B' < A'B'$, and so we conclude that

$$\frac{D'B'}{B'C'} < \frac{A'B'}{B'C'}, \qquad \frac{m}{n} < \frac{A'B'}{B'C'}.$$

COROLLARY TO THEOREM 3.2. *If* $\dfrac{m}{n} > \dfrac{AB}{BC}$, *then* $\dfrac{m}{n} > \dfrac{A'B'}{B'C'}$.

This corollary is analogous to Theorem 3.2, and so a similar proof holds.

Thus we have proved Theorem 3.2 and a corollary; we shall now use these to prove Theorem 3.1 in the irrational case. Let β denote the irrational number which represents the ratio AB/BC. We use the decimal representation of β as in Section 3.2.

To illustrate what we are about to do, let β have the value $\pi = 3.14159 \cdots$ for example. Then we can write

$$\frac{3}{1} < \beta < \frac{4}{1},$$

$$\frac{31}{10} < \beta < \frac{32}{10},$$

(1)

$$\frac{314}{100} < \beta < \frac{315}{100},$$

$$\frac{3141}{1000} < \beta < \frac{3142}{1000}, \ldots, \text{etc.}$$

The fractions on the left are obtained by taking the numbers 3, 3.1, 3.14, 3.141, etc. from the decimal expansion. The fractions on the right are obtained by increasing these same numbers by 1, 0.1, 0.01, 0.001, etc.

The chain of inequalities (1) is infinite in number; we have written only the first four. These inequalities *characterize* the particular value of β we are discussing, namely π. That is to say, if a number β satisfies all the inequalities (1), then that number equals π.

The inequalities (1) were written in connection with an illustrative example, where β had the value π. Now we drop this example, but we point out that whatever irrational value β has, its decimal representation will provide us with a chain of inequalities

$$\frac{a_1}{1} < \beta < \frac{1 + a_1}{1},$$

$$\frac{a_2}{10} < \beta < \frac{1 + a_2}{10},$$

(2)

$$\frac{a_3}{100} < \beta < \frac{1 + a_3}{100},$$

$$\frac{a_4}{1000} < \beta < \frac{1 + a_4}{1000}, \ldots, \text{etc.}$$

which characterizes β uniquely, and in each inequality β is between two rational numbers. The symbols a_1, a_2, a_3, $\ldots$ denote integers.

Our plan is to let β' denote the ratio $A'B'/B'C'$ and to prove that β' also satisfies the inequalities (2), just as β does. But these inequalities characterize the number β, and hence β' is identified with β, so that

$$\beta = \frac{AB}{BC} = \frac{A'B'}{B'C'} = \beta'.$$

All that remains is to show that β' satisfies all the inequalities (2). To do this we use Theorem 3.2. First we take any of the fractions $a_1/1$, $a_2/10$, $a_3/100$, etc., say $a_3/100$, and interpret this as the rational number m/n of Theorem 3.2. Then the hypothesis of Theorem 3.2,

$$\frac{m}{n} < \frac{AB}{BC},$$

becomes

$$\frac{a_3}{100} < \beta,$$

and this is valid because of the inequalities (2). Then Theorem 3.2 tells us that

$$\frac{m}{n} < \frac{A'B'}{B'C'};$$

that is,

$$\frac{a_3}{100} < \beta'.$$

Thus we see that β' satisfies

$$\frac{a_1}{1} < \beta', \qquad \frac{a_2}{10} < \beta', \qquad \frac{a_3}{100} < \beta', \qquad \frac{a_4}{1000} < \beta', \qquad \text{etc.}$$

By an analogous use of the corollary to Theorem 3.2, we get the inequalities

$$\beta' < \frac{1 + a_1}{1}, \qquad \beta' < \frac{1 + a_2}{10}, \qquad \beta' < \frac{1 + a_3}{100}, \qquad \beta' < \frac{1 + a_4}{1000}, \qquad \text{etc.}$$

Hence β' satisfies the inequalities (2) just as β does. Thus $\beta = \beta'$ and the proof of Theorem 3.1 is complete.

3.8 A Summary

In this chapter, we have indicated that every real number can be associated with exactly one point on the "real line." We have also seen that every real number has exactly one representation as an infinite decimal (provided we exclude an infinite succession of zeros, i.e., terminating decimals, in our representations). This representation by infinite decimals was applied in Section 3.7 to a key theorem in elementary geometry. In addition we have demonstrated the irrationality of certain numbers such as $\sqrt{2}$, $\sqrt{3}$, $\sqrt{2} + \sqrt{3}$, etc. Our methods, however, were rather piecemeal, and we did not give any very general procedure for determining whether or not a given number is rational.

In the next chapter, we shall study irrational numbers in a much more systematic way. We shall derive a method by which a large class of numbers can be classified as irrational.

CHAPTER FOUR

Irrational Numbers

In the course of this chapter and the next, we shall learn that the real numbers can be classified not only into rational and irrational numbers, but also into two other categories. One category contains the so-called *algebraic numbers*, i.e., those numbers which are solutions of algebraic equations with integer coefficients, and the other includes all remaining numbers and these are called *transcendental numbers*. This distinction will become more meaningful in what follows. We mention at once, however, that some algebraic numbers are rational and some are irrational, but all transcendental numbers are irrational.

The over-all purpose of this chapter is to devise a systematic method for determining whether or not a given algebraic number is rational. (Actually, we shall not treat the class of algebraic numbers in its greatest generality, but we shall apply our method to many examples.) But before we derive this method, we shall study some simple properties of irrational numbers.

4.1 Closure Properties

In contrast to the rational numbers which were shown to be closed under addition, subtraction, multiplication, and division (except by zero), the irrational numbers possess none of these properties. Before showing this, we prove a theorem which will enable us to manufacture infinitely many irrational numbers from one given irrational number.

THEOREM 4.1. *Let α be any irrational number and r any rational number except zero. Then the addition, subtraction, multiplication, and division of r and α yield irrational numbers. Also $-\alpha$ and α^{-1} are irrational.*

PROOF. These results are readily established by means of indirect proofs. Suppose, to begin with, that $-\alpha$ were rational, say $-\alpha = r'$, where r' denotes the presumed rational number. Then we would have $\alpha = -r'$, and $-r'$ is also a rational number. Thus we have a contradiction because α is irrational.

The theorem asserts that $-\alpha$, $\alpha^{-1} = 1/\alpha$, $\alpha + r$, $\alpha - r$, $r - \alpha$, $r\alpha$, α/r, and r/α are irrational. We have already treated $-\alpha$. In order to prove the irrationality of α^{-1}, we observe that it is a special case of r/α with $r = 1$. Thus there is no need to treat this case separately.

Let us prove the remaining six cases all at once, in wholesale fashion. If one or more of these expressions were rational, then one or more of the following equations,

$$\alpha + r = r_1, \quad \alpha - r = r_2, \quad r - \alpha = r_3, \quad r\alpha = r_4, \quad \frac{\alpha}{r} = r_5, \quad \frac{r}{\alpha} = r_6,$$

would hold, where $r_1, r_2, r_3, r_4, r_5, r_6$ denote some rational numbers. Solving these equations for α, we get

$$\alpha = r_1 - r, \quad \alpha = r_2 + r, \quad \alpha = r - r_3, \quad \alpha = \frac{r_4}{r}, \quad \alpha = rr_5, \quad \alpha = \frac{r}{r_6}.$$

The right-hand members of these equations are rational numbers because of the closure properties of rational numbers. But none of these equations holds, since α is irrational. Hence it is not possible for any of the numbers $\alpha + r$, $\alpha - r$, etc. to be rational. Thus the proof of the theorem is complete.

By means of Theorem 4.1, we can construct a large class of irrational numbers from one single irrational number, for example, from $\sqrt{2}$. Applying each statement of the theorem, we can assert, for example, that

$$-\sqrt{2}, \quad \frac{1}{\sqrt{2}}, \quad \sqrt{2} + 5, \quad 3 - \sqrt{2}, \quad -2\sqrt{2}, \quad \frac{\sqrt{2}}{7}, \quad \frac{4}{\sqrt{2}}$$

are all irrational. Since there are infinitely many rational numbers which we can use in each of the assertions of the theorem, it is clear that infinitely many irrational numbers can thus be manufactured.

Moreover, any one of the numbers so constructed, say for example $\sqrt{2} + 5$, can now be used as a new irrational number α in the theorem. Therefore, infinitely many more irrational numbers, e.g.,

$$-\sqrt{2} - 5, \quad \frac{1}{\sqrt{2} + 5}, \quad \sqrt{2} + 8, \quad 5\sqrt{2} + 25, \quad \frac{\sqrt{2} + 5}{7}, \quad \text{etc.}$$

can be generated from that one number.

Are the irrational numbers closed under addition? No, they are not.

To prove this we need only exhibit two irrational numbers whose sum is rational. Now, in the preceding chapter $\sqrt{2}$ was shown to be irrational, therefore $-\sqrt{2}$ is irrational by Theorem 4.1. But the sum of $\sqrt{2}$ and $-\sqrt{2}$ is 0, which is rational; so is the sum of $3 + \sqrt{2}$ and $5 - \sqrt{2}$, for example. More generally, the sum of $r_1 + \alpha$ and $r_2 - \alpha$ (where r_1 and r_2 are rational and α is irrational) is rational.

The proposition that the irrational numbers are not closed under addition does not mean that if we add *any two* irrational numbers the sum will be rational. It means only that there is at least one case where the sum *is* rational. The result obtained when two irrational numbers are added may be either rational or irrational, depending on the two numbers we start with. Whereas the sum of $\sqrt{2}$ and $-\sqrt{2}$ is a rational number, the sum of $\sqrt{2}$ and $\sqrt{3}$ is an irrational number, as we established in the preceding chapter.

Are the irrational numbers closed under subtraction? No, because, for example, if we subtract $\sqrt{2}$ from itself we get the rational number 0.

Similarly, the irrational numbers are not closed under multiplication or division. These results are so similar to the earlier ones that we leave the proofs to the reader in the next set of problems.

Problem Set 11

(In some of these problems, it may be helpful to use some of the results derived in the preceding chapter, namely that $\sqrt{2}$, $\sqrt{3}$, $\sqrt{6}$, and $\sqrt{2} + \sqrt{3}$ are irrational.)

1. Exhibit two irrational numbers whose difference is irrational.
2. Exhibit two irrational numbers whose product is rational and, so prove that the irrational numbers are not closed under multiplication.
3. Exhibit two irrational numbers whose product is irrational.
4. Exhibit two irrational numbers whose quotient is rational, and so prove that irrational numbers are not closed under division.
5. Exhibit two irrational numbers whose quotient is irrational.
*6. Prove that $\sqrt{3}\,(\sqrt{6} - 3)$ is irrational.
7. Let α be a positive irrational number. Prove that $\sqrt{\alpha}$ is irrational.
8. Given that α and β are irrational, but $\alpha + \beta$ is rational, prove that $\alpha - \beta$ and $\alpha + 2\beta$ are irrational.

4.2 Polynomial Equations

It was proved in the preceding chapter that $\sqrt{2}$, $\sqrt{3}$, and $\sqrt{6}$ are irrational. As might be expected (or, perhaps, as the reader already knows) such numbers as $\sqrt{7}$, $\sqrt[3]{5}$, and $\sqrt[5]{91}$ are also irrational. What we want to do next is establish the irrationality of all such numbers by a common scheme instead of treating them one at a time. To do this, we shall shift the emphasis from the numbers themselves to simple algebraic equations having the numbers as roots. For example, $\sqrt{2}$ is a root of the equation $x^2 - 2 = 0$; other ways of saying this are: "$\sqrt{2}$ is a solution of $x^2 - 2 = 0$" or "$\sqrt{2}$ satisfies the equation $x^2 - 2 = 0$." Similarly, the other numbers under consideration satisfy equations as follows:

$$\sqrt{3}, \quad x^2 - 3 = 0,$$

$$\sqrt{6}, \quad x^2 - 6 = 0,$$

$$\sqrt{7}, \quad x^2 - 7 = 0,$$

$$\sqrt[3]{5}, \quad x^3 - 5 = 0,$$

$$\sqrt[5]{91}, \quad x^5 - 91 = 0.$$

What we shall do is to establish that these equations, and more generally all equations satisfying certain conditions, have no rational roots. To begin we must define a few terms used to describe equations.

By a *quadratic polynomial* in x, we mean an expression of the form $ax^2 + bx + c$, where a, b, and c are called the coefficients. A *cubic polynomial*, or a *polynomial of degree* 3, is of the form $ax^3 + bx^2 + cx + d$. In order to avoid introducing new letters as the degree is increased, it is convenient to write

$$c_3x^3 + c_2x^2 + c_1x + c_0.$$

Now, a polynomial of any degree n (where n is a positive integer) has the form

$$c_nx^n + c_{n-1}x^{n-1} + \cdots + c_1x + c_0, \quad \text{with } c_n \text{ not zero.}$$

A *polynomial equation* is a statement of equality of the form

(1) $$c_nx^n + c_{n-1}x^{n-1} + \cdots + c_1x + c_0 = 0;$$

$c_0, c_1, c_2, \ldots, c_n$ are called its *coefficients*.

EXAMPLE. Identify the values of n, c_n, etc., when the equation
$$3x^6 + 2x^5 - x^4 + 10x^3 + 4x - 7 = 0$$
is thought of as having the form (1) above.

SOLUTION. By direct comparison we see that

$$n = 6, \quad c_6 = 3, \quad c_5 = 2, \quad c_4 = -1, \quad c_3 = 10, \quad c_2 = 0, \quad c_1 = 4, \quad c_0 = -7.$$

Observe that the requirement that the coefficients of eq. (1) be integers is no stricter than the requirement that the coefficients be rational; for, if they are rational then $c_0 = a_0/b_0$, $c_1 = a_1/b_1$, $c_2 = a_2/b_2$, ..., where the a's and b's are integers. All these fractions can be written so as to have a common denominator, for example the product $b_0 b_1 b_2 \cdots b_n$, by which we may multiply both sides of the equation and thus obtain a new equation whose coefficients are integers and whose roots are the same as those of the original equation.

We recall that a *root* of an equation in x is a value which, when substituted for x, satisfies the equation. For example, as we noted earlier, $\sqrt{7}$ is a root of $x^2 - 7 = 0$.

EXAMPLE. Is 2/5 a root of $10x^3 + 6x^2 + x - 2 = 0$?

SOLUTION. By substitution of 2/5 for x, we get

$$10\left(\frac{2}{5}\right)^3 + 6\left(\frac{2}{5}\right)^2 + \frac{2}{5} - 2 = 0,$$

and this is a correct statement of arithmetic. Therefore 2/5 is a root of the equation.

We are now ready to return to the main point. We repeat that the method we are about to develop for deciding whether or not a given number is irrational can be applied if and only if we can write down a polynomial equation which has the number under consideration as a root. The method can be used not only for the numbers whose irrationality we established in the preceding chapter, but also for any number which can be written as a finite combination of the symbols $+$, $-$, $\times$, $\div$, and radicals $\sqrt[n]{r}$ of rational numbers. For example,

$$\frac{\sqrt{\sqrt{2} + \sqrt{3}} + \sqrt[3]{\sqrt[10]{7} - \sqrt[11]{7}}}{\sqrt[156]{25}}$$

is a complicated case of the kind of number we are talking about.

We do not prove in this book that all such numbers are roots of polynomial equations with integer coefficients, but we shall write the polynomial equations satisfied by many such numbers.

Problem Set 12

1. Identify the values of n, c_n, etc., when the following equations are thought of as having the form of eq. (1) above:

(a) $15x^3 - 23x^2 + 9x - 1 = 0$;
(b) $3x^3 + 2x^2 - 3x - 2 = 0$;
(c) $2x^3 + 7x^2 - 3x - 18 = 0$;
(d) $2x^4 - x^2 - 3x + 5 = 0$;
(e) $3x^5 - 5x^3 + 6x^2 - 12x + 8 = 0$;
(f) $x^4 - 3x^2 - 5x + 9 = 0$.

2. (a) Is $1/3$ a root of (a) above?
   (b) Is $-2/3$ a root of (b) above?
   (c) Is $3/2$ a root of (c) above?
   (d) Is 2 a root of (d) above?
   (e) Is -2 a root of (e) above?
   (f) Is $1/2$ a root of (f) above?

3. Establish that $\sqrt{7}$ is a root of $\frac{1}{3} x^2 - \frac{7}{3} = 0$.

4. Prove that if a number is a root of a polynomial equation such as

$$\frac{a_3}{b_3} x^3 + \frac{a_2}{b_2} x^2 + \frac{a_1}{b_1} x + \frac{a_0}{b_0} = 0$$

with rational coefficients a_3/b_3, etc., then that number is a root of a polynomial equation with integer coefficients.

5. Generalize the result of the preceding problem from equations of degree 3 to equations of degree n.

4.3 Rational Roots of Polynomial Equations

Our purpose now is to derive a simple rule, given as Theorem 4.3 below, which enables us to find all rational roots of any given polynomial equation with integer coefficients. Thus we will be able to separate the rational roots and the irrational roots of an equation, and so establish the irrationality of a wide class of numbers.

But first we need the following auxiliary result.

THEOREM 4.2. *Let u, v, w be integers such that u is a divisor of vw, and u and v have no prime factor in common. Then u is a divisor of w. More generally, if u is a divisor of $v^n w$, where n is any positive integer and u and v have no prime factor in common, then u is a divisor of w.*

Before presenting the proof, let us illustrate this proposition with some examples.

(1) Let $u = 2$, $v = 3$, and $vw = 12$. Now, 2 and 3 have no prime factor in common. Also 2 is a divisor of 12, so the hypothesis of Theorem 4.2 is satisfied. The conclusion that 2 is a divisor of $w = 12/v = 4$ is valid.

(2) Let $u = 4$, $v = 5$, $v^3 w = 500$. The numbers 4 and 5 have no prime factor in common and 4 divides 500. The conclusion of the more general statement, namely, that 4 divides $w = 500/125 = 4$, also holds.

PROOF. The main ingredient in this proof is the Fundamental Theorem of Arithmetic which is proved in Appendix B at the end of the book and which assures us that there is only one way to factor u, v, and w into prime factors. Since u divides vw, all the prime factors of u occur also in vw; moreover, if any prime p occurs in u to the power α, then it occurs in vw to at least that same power, i.e., it occurs in vw to a power β, where $\beta \geq \alpha$. Now, since u and v have no prime factor in common, it follows that all prime factors of u occur, to at least the same power, in the factoring of w. Hence u is a divisor of w.

The last statement in the theorem can be argued in a similar way. The assumption that u and v have no common prime factor assures us that u and v^n have no prime factor in common. Once again it follows that v^n in no way contributes to the fact that u is a divisor of $v^n w$, and so u must be a divisor of w.

We now have enough background material to state and prove the following proposition.

THEOREM 4.3. *Consider any polynomial equation with integer coefficients,*

(1) $$c_n x^n + c_{n-1} x^{n-1} + c_{n-2} x^{n-2} + \cdots + c_2 x^2 + c_1 x + c_0 = 0.$$

If this equation has a rational root a/b, where a/b is presumed to be in lowest terms, then a is a divisor of c_0 and b is a divisor of c_n.

Again we shall illustrate this statement by an example before we present its proof. Consider the equation

$$2x^3 - 9x^2 + 10x - 3 = 0.$$

The theorem says that, if a/b is a rational root in lowest terms, then a is a divisor of -3 and b is a divisor of 2. Hence, the possible values for a are $+1$, -1, $+3$, -3, and those of b are $+1$, -1, $+2$, -2. Combining these possibilities, we find that the following set contains all possible roots:

$$\frac{+1}{+1}, \ \frac{+1}{-1}, \ \frac{+1}{+2}, \ \frac{+1}{-2}, \ \frac{-1}{+1}, \ \frac{-1}{-1}, \ \frac{-1}{+2}, \ \frac{-1}{-2},$$

$$\frac{+3}{+1}, \ \frac{+3}{-1}, \ \frac{+3}{+2}, \ \frac{+3}{-2}, \ \frac{-3}{+1}, \ \frac{-3}{-1}, \ \frac{-3}{+2}, \ \frac{-3}{-2}.$$

This list contains only eight distinct numbers, namely, 1, -1, 1/2, $-1/2$, 3, -3, 3/2, $-3/2$. Of these, only the numbers 1, 1/2, and 3 are actual roots of the equation, as the reader may verify by substitution.

PROOF. Let a/b be a root of eq. (1). This means that if a/b is substituted for x, then

$$(2) \qquad c_n\left(\frac{a}{b}\right)^n + c_{n-1}\left(\frac{a}{b}\right)^{n-1} + \cdots + c_2\left(\frac{a}{b}\right)^2 + c_1\left(\frac{a}{b}\right) + c_0 = 0$$

is true. We begin by giving the proof for the special case where $n = 3$, because it will be easier for the reader to follow. Then we shall give an analogous argument for the general case.

In the case $n = 3$, eq. (2) is simply

$$c_3\left(\frac{a}{b}\right)^3 + c_2\left(\frac{a}{b}\right)^2 + c_1\left(\frac{a}{b}\right) + c_0 = 0.$$

Multiplying by b^3, we get

$$(3) \qquad c_3 a^3 + c_2 a^2 b + c_1 a b^2 + c_0 b^3 = 0.$$

First we write eq. (3) in the form

$$c_3 a^3 = -c_2 a^2 b - c_1 a b^2 - c_0 b^3$$

or

$$c_3 a^3 = b(-c_2 a^2 - c_1 a b - c_0 b^2).$$

This shows that b is a divisor of $c_3 a^3$. We apply Theorem 4.2 at this point, with u, v, and w replaced by b, a, and c_3, respectively. The hypothesis of Theorem 4.2, that u and v have no prime factor in common, is satisfied because a/b is in lowest terms so that a and b have no prime factor in common. Hence we can conclude from Theorem 4.2 that b is a divisor of c_3. This is a part of the desired conclusion in Theorem 4.3, because in this case $n = 3$, so that c_n is c_3.

Next we write eq. (3) in the form

$$c_0 b^3 = -c_1 a b^2 - c_2 a^2 b - c_3 a^3$$

or

$$c_0 b^3 = a(-c_1 b^2 - c_2 a b - c_3 a^2).$$

This shows that a is a divisor of $c_0 b^3$. By an argument virtually identical to the earlier one, i.e., by again applying Theorem 4.2, we conclude that a is a divisor of c_0. Thus the proof is complete in case $n = 3$.

To prove the theorem for any n, we return to eq. (2) and multiply it by b^n to obtain

(4) $c_n a^n + c_{n-1} a^{n-1} b + \cdots + c_2 a^2 b^{n-2} + c_1 a b^{n-1} + c_0 b^n = 0.$

Now (4) can be rewritten as

$$c_n a^n = - c_{n-1} a^{n-1} b - \cdots - c_2 a^2 b^{n-2} - c_1 a b^{n-1} - c_0 b^n$$

or

$$c_n a^n = b(- c_{n-1} a^{n-1} - \cdots - c_2 a^2 b^{n-3} - c_1 a b^{n-2} - c_0 b^{n-1}).$$

This shows that b is a divisor of $c_n a^n$. We apply Theorem 4.2 with u, v, and w replaced by b, a, and c_n, respectively, and conclude that b is a divisor of c_n.

Next we rewrite eq. (4) as

$$c_0 b^n = a(-c_n a^{n-1} - \cdots - c_2 a b^{n-2} - c_1 b^{n-1}).$$

This shows that a is a divisor of $c_0 b^n$. Again by applying Theorem 4.2 with u, v, w replaced by a, b, c_0, respectively, we conclude that a is a divisor of c_0. This completes the proof of Theorem 4.3.

We could have avoided the argument of the last paragraph by observing that there is a symmetry about eq. (4), and that b stands in relation to c_n in the equation exactly as a stands in relation to c_0.

Next we examine the situation that holds when $c_n = 1$.

COROLLARY 1. *Consider an equation of the form*

$$x^n + c_{n-1} x^{n-1} + c_{n-2} x^{n-2} + \cdots + c_2 x^2 + c_1 x + c_0 = 0,$$

whose coefficients are integers. If such an equation has a rational root, it is an integer; moreover, this integer root is a divisor of c_0.

PROOF. Consider any rational root a/b. We may presume that b is a positive integer, because if b were negative we could absorb the minus sign in a. According to Theorem 4.3, b must be a divisor of c_n; that is, b must be a divisor of 1. But $+1$ and -1 are the only divisors of 1, and so we must have $b = +1$, since we ruled out any negative value for b. Consequently any rational root is of the form $a/1$, and so it is an integer a. Also by Theorem 4.3 we know that a is a divisor of c_0, and thus the proof of the corollary is complete.

EXAMPLE. Prove that $\sqrt{7}$ is irrational.

SOLUTION. $\sqrt{7}$ is a root of $x^2 - 7 = 0$. Here, according to our notation, $c_2 = 1$ and $c_0 = -7$.

Now, there are two ways we can proceed. One way is to use Corollary 1 and say: If $x^2 - 7 = 0$ has a rational root a/b, then that rational root would have to be an integer. We can show that $\sqrt{7}$ is not an integer and that therefore it is not a rational root of $x^2 - 7 = 0$. Hence it must be an irrational root. Clearly, $\sqrt{7}$ is not an integer because it lies between the consecutive integers 2 and 3; this, in turn, follows from the inequalities

$$4 < \phantom{\sqrt{}}7 < \phantom{\sqrt{}}9,$$
$$\sqrt{4} < \sqrt{7} < \sqrt{9},$$
$$2 < \sqrt{7} < \phantom{\sqrt{}}3.$$

Another way employs Corollary 1 in its full form, according to which any rational root of $x^2 - 7 = 0$ is an integer which is an exact divisor of -7. The only divisors of -7 are 1, -1, 7, and -7. But none of these is a root as can be seen by simple verification; the equations

$$1^2 - 7 = 0, \qquad (-1)^2 - 7 = 0, \qquad 7^2 - 7 = 0, \qquad (-7)^2 - 7 = 0.$$

are all false. Hence $x^2 - 7 = 0$ has no integral root, hence no rational root, and $\sqrt{7}$ is an irrational number.

EXAMPLE. Prove that $\sqrt[3]{5}$ is irrational.

SOLUTION. $\sqrt[3]{5}$ is a root of $x^3 - 5 = 0$. According to Corollary 1, if this equation has a rational root it is an integer which is a divisor of 5. The divisors of 5 are $+1$, -1, $+5$, and -5. But none of these is a root because the equations

$$1^3 - 5 = 0, \qquad (-1)^3 - 5 = 0, \qquad 5^3 - 5 = 0, \qquad (-5)^3 - 5 = 0$$

are all false. Hence $x^3 - 5 = 0$ has no rational roots, and so $\sqrt[3]{5}$ is irrational.

These two examples are special cases of the following more general result:

COROLLARY 2. *A number of the form $\sqrt[n]{a}$, where a and n are positive integers, is either irrational or it is an integer; in the latter case a is the nth power of an integer.*

PROOF. This follows from Corollary 1 because $\sqrt[n]{a}$ is a root of $x^n - a = 0$, and if this equation has a rational root, it must be an integer. Furthermore, if $\sqrt[n]{a}$ is an integer, say k, then $a = k^n$.

Problem Set 13

1. Prove that $\sqrt{2}$, $\sqrt{3}$, $\sqrt{13}$, and $\sqrt[5]{91}$ are irrational.
2. Prove that $(4\sqrt{13} - 3)/6$ is irrational.

3. Prove that $\sqrt{15}$ is irrational.

4. Prove that $4/(16 - 3\sqrt{15})$ is irrational.

5. Prove that $\sqrt[3]{6}$ is irrational.

6. Prove that $(1/3)(2\sqrt[3]{6} + 7)$ is irrational.

7. Prove that Theorem 4.3 becomes a false statement if the words "presuming that a/b is in lowest terms" are omitted.

4.4 Further Examples

In Chapter 3, it was proved by a method which applies to a rather wide class of numbers that $\sqrt{2} + \sqrt{3}$ is irrational. However, an even wider class of numbers can be treated with the aid of Corollary 1.

Let us discuss $\sqrt{2} + \sqrt{3}$ again. If we write $x = \sqrt{2} + \sqrt{3}$, then we have

$$x - \sqrt{2} = \sqrt{3}.$$

Squaring both sides, we obtain

$$x^2 - 2x\sqrt{2} + 2 = 3,$$

and by rearranging terms, we have

$$x^2 - 1 = 2x\sqrt{2}.$$

If we square this again, we obtain

$$x^4 - 2x^2 + 1 = 8x^2$$

or

(5) $$x^4 - 10x^2 + 1 = 0.$$

Because of the way in which eq. (5) has been constructed, we know that $\sqrt{2} + \sqrt{3}$ is a root. Next, we shall apply Corollary 1 to show that eq. (5) has no rational roots, and from this we shall conclude that $\sqrt{2} + \sqrt{3}$ is irrational.

The application of Corollary 1 to eq. (5) tells us that if this equation has any rational roots, they must be integers which divide 1. But the only divisors of 1 are $+1$ and -1, neither of which is a root of $x^4 - 10x^2 + 1 = 0$. We thus conclude that eq. (5) has no rational roots and that $\sqrt{2} + \sqrt{3}$ is irrational.

Another way of reaching the same conclusion is this: Instead of testing whether $+1$ and -1 are roots of eq. (5), we may argue as follows.

Even if $+1$ or -1, or both, were roots of eq. (5), we can observe that the root $\sqrt{2} + \sqrt{3}$ is different from $+1$ and from -1; for example, we can argue that both $\sqrt{2}$ and $\sqrt{3}$ are larger than 1, so that their sum is too large to be equal to $+1$ or -1. Hence $\sqrt{2} + \sqrt{3}$ is not among the possible rational roots of eq. (5), regardless of whether or not $+1$ or -1 are actual roots. It follows that $\sqrt{2} + \sqrt{3}$ is irrational.

EXAMPLE. Prove that $\sqrt[3]{2} - \sqrt{3}$ is irrational.

SOLUTION. Writing $x = \sqrt[3]{2} - \sqrt{3}$, we see that

$$x + \sqrt{3} = \sqrt[3]{2}.$$

Now, cube both sides and obtain

$$x^3 + 3\sqrt{3}x^2 + 9x + 3\sqrt{3} = 2.$$

When the terms are rearranged,

$$x^3 + 9x - 2 = -3\sqrt{3}(x^2 + 1).$$

By squaring, we get

$$x^6 + 18x^4 - 4x^3 + 81x^2 - 36x + 4 = 27\,(x^4 + 2x^2 + 1)$$

or

$$x^6 - 9x^4 - 4x^3 + 27x^2 - 36x - 23 = 0.$$

This equation has been so constructed that $\sqrt[3]{2} - \sqrt{3}$ is a root. But the only possible rational roots of this equation are integers which are divisors of -23. Hence the only possible rational roots are $+1$, -1, $+23$, and -23, and these are not roots, as straightforward substitution shows:

$+1$: $1^6 - 9(1)^4 - 4(1)^3 + 23(1)^2 - 36(1) - 23 = 0$ (False!)

-1: $(-1)^6 - 9(-1)^4 - 4(-1)^3 + 27(-1)^2 - 36(-1) - 23 = 0$ (False!)

23: $(23)^6 - 9(23)^4 - 4(23)^3 + 27(23)^2 - 36(23) - 23 = 0$

       (False, because for example, $(23)^6$ is much too large to be canceled out by the other terms!)

-23: $(-23)^6 - 9(-23)^4 - 4(-23)^3 + 27(-23)^2 - 36(-23) - 23 = 0$ (False!)

Hence there are no rational roots and so $\sqrt[3]{2} - \sqrt{3}$ is irrational.

As in the preceding example, there is no need to test whether $+1, -1, +23$, and -23 are roots of the equation. Instead, we can argue that $\sqrt[3]{2} - \sqrt{3}$ is different from any of these four possible rational roots. We observe that $\sqrt[3]{2}$ is in the vicinity of 1.2 and $\sqrt{3}$ is in the vicinity of 1.7. Consequently $\sqrt[3]{2} - \sqrt{3}$ is approximately -0.5, and hence not equal to any of the values $+1, -1, +23$, or -23. It follows that the root $\sqrt[3]{2} - \sqrt{3}$ is irrational, since it is different from all the possible rational roots.

Problem Set 14

1. Prove that $\sqrt{3} - \sqrt{2}$ is irrational.

2. Prove that $\sqrt[3]{3} + \sqrt{2}$ is irrational.

3. Prove that $\sqrt[3]{5} - \sqrt{3}$ is irrational.

4.5 A Summary

In this chapter we have been dealing with so-called "algebraic irrationalities." We have seen that there are infinitely many irrational numbers and we have studied ways of constructing some of them from a given irrational number.

We have also found the following method for testing whether or not a given number k is irrational:

First we look for a polynomial equation

$$c_n x^n + c_{n-1} x^{n-1} + \cdots + c_1 x + c_0 = 0$$

satisfied by the value $x = k$. (If we cannot find such an equation, we cannot apply this method.)

Next we apply Theorem 4.3, or, if $c_n = 1$, Corollary 1. Often it is clear that the equation possesses no rational roots at all. Then k is clearly an irrational root. Sometimes, we see at a glance that k is different from all the possible candidates for rational roots of the equation, and so we may deduce the irrationality of k. Or, by direct substitution, we select from all the candidates those rational numbers which are actual roots of the equation. Then, in order to prove that k is irrational, we must show that k is different from all the rational roots.

In the next chapter we shall use the methods of this chapter to demonstrate the irrationality of many trigonometric numbers, and we shall use the Fundamental Theorem of Arithmetic to demonstrate the irrationality of many logarithmic numbers. Moreover, we shall learn that there are irrational numbers which are not roots of algebraic equations.

CHAPTER FIVE

Trigonometric and Logarithmic Numbers

The reader is undoubtedly familiar† with such trigonometric functions as $\sin \theta$ and $\cos \theta$, and knows that each of these functions assigns a real number to every angle θ. He has probably also encountered the logarithmic function $\log x$ which assigns a real number to every positive real number x.

If θ measured in degrees is rational, then the trigonometric functions of θ are irrational, apart from a few exceptions; similarly, if x is rational then $\log x$ is irrational, apart from a few special cases.‡

Although our attention is confined to certain simple examples in this chapter, a much stronger result is obtained in Appendix D by a more advanced and more difficult method.

5.1 Irrational Values of Trigonometric Functions

We shall show, using the methods of the previous chapter and certain basic trigonometric identities, that for many angles θ the corresponding values of the trigonometric functions are irrational.

To this end, let us first recall the following basic trigonometric formulas:

(1) $$\cos (A + B) = \cos A \cos B - \sin A \sin B,$$

(2) $$\sin (A + B) = \sin A \cos B + \cos A \sin B.$$

† Readers who have not yet studied trigonometry or logarithms can find an introduction to these subjects in *Plane Trigonometry* by A. L. Nelson and K. W. Folley, Harper, 1956.

‡ Tables listing numbers in decimal representations must restrict these to a finite number of places; thus irrational numbers are approximated in such tables.

Replacing A and B by a single value, say θ, we get

(3)
$$\cos 2\theta = \cos^2\theta - \sin^2\theta,$$

(4)
$$\sin 2\theta = 2 \sin\theta \cos\theta.$$

Next, if we replace A by 2θ and B by θ in (1), we get

$$\cos 3\theta = \cos 2\theta \cos \theta - \sin 2\theta \sin \theta$$

Using (3) and (4) and also the well-known identity $\cos^2\theta + \sin^2\theta = 1$, we obtain

$$\cos 3\theta = (\cos^2\theta - \sin^2\theta) \cos\theta - (2 \sin\theta \cos\theta) \sin\theta,$$
$$= \cos^3\theta - 3 \sin^2\theta \cos\theta,$$
$$= \cos^3\theta - 3(1 - \cos^2\theta) \cos\theta$$

or

(5)
$$\cos 3\theta = 4 \cos^3\theta - 3 \cos\theta.$$

Now consider the number $\cos 20°$. By setting $\theta = 20°$ in (5), we have

$$\cos 60° = 4 \cos^3 20° - 3 \cos 20°.$$

If we write x for $\cos 20°$, and make use of the fact that $\cos 60° = \frac{1}{2}$, we get

$$\frac{1}{2} = 4x^3 - 3x$$

or

(6)
$$8x^3 - 6x - 1 = 0.$$

Because of the construction of eq. (6), we know that $\cos 20°$ is a root. Applying Theorem 4.3 to eq. (6), we see that the only possible rational roots of this equation are ± 1, $\pm\frac{1}{2}$, $\pm\frac{1}{4}$, $\pm\frac{1}{8}$. But none of these eight possibilities is an actual root, as can be seen by substitution into eq. (6). Hence we conclude that eq. (6) has no rational roots and so $\cos 20°$ is an irrational number.

This conclusion can also be reached without testing whether the possible rational roots ± 1, $\pm\frac{1}{2}$, $\pm\frac{1}{4}$, $\pm\frac{1}{8}$ are actual roots of eq. (6). It is enough to show that cos 20° is different from all these eight values. This we can do by looking at the value given for cos 20° in a table of trigonometric functions. (Such a table gives an approximation only, of course.) Or, we can observe that cos 20° lies between cos 0° and cos 30°, the cosine being a decreasing function for these angles. Thus we see that cos 20° lies between 1 and $\sqrt{3}/2$; thus it is between 1 and 0.8. It follows that cos 20° cannot be equal to any of the possible rational roots of eq. (6), and therefore cos 20° is an irrational number.

EXAMPLE. Prove that sin 10° is irrational.

FIRST SOLUTION. One way to solve this problem would be to begin with the trigonometric identity for sin 3θ, i.e.,

$$(7) \qquad \sin 3\theta = 3 \sin \theta - 4 \sin^3\theta,$$

which can be obtained from (2) in the same way that (5) was obtained from (1). Replacing θ by 10° in (7), and using the fact that sin 30° = 1/2, we get

$$\frac{1}{2} = 3 \sin 10° - 4 \sin^3 10°.$$

Write x for sin 10° to obtain

$$\frac{1}{2} = 3x - 4x^3, \qquad 8x^3 - 6x + 1 = 0.$$

As with eq. (6), it is not difficult to show (by Theorem 4.3) that the equation $8x^3 - 6x + 1 = 0$ has no rational roots. Hence sin 10° is irrational.

SECOND SOLUTION. Eq. (3) has two alternative forms,

$$(8) \qquad \cos 2\theta = 2 \cos^2\theta - 1, \qquad \cos 2\theta = 1 - 2 \sin^2\theta,$$

both of which can be obtained from (3) by use of the basic identity

$$\sin^2\theta + \cos^2\theta = 1.$$

If, in the second alternative of (8), we replace θ by 10°, we get

$$(9) \qquad \cos 20° = 1 - 2 \sin^2 10°.$$

Now suppose that sin 10° were rational. Then both $\sin^2 10°$ and $1 - 2 \sin^2 10°$ also would be rational. But cos 20° is irrational, as we have already proved. Hence we have a contradiction, so sin 10° is irrational.

Problem Set 15

In solving these problems, make use (wherever helpful) of any previous results, either in the text or in the problems themselves.

1. Prove that the following numbers are irrational:

    (a) cos 40°   (b) sin 20°   (c) cos 10°   (d) sin 50°

2. Establish identity (7).

3. (a) Establish the identity $\cos 5\theta = 16\cos^5\theta - 20\cos^3\theta + 5\cos\theta$.
   (b) Prove that $\cos 12°$ is irrational.

4. Which of the following are rational?

(a)	sin 0°	(d)	sin 30°	(g)	sin 45°	(j)	sin 60°
(b)	cos 0°	(e)	cos 30°	(h)	cos 45°	(k)	cos 60°
(c)	tan 0°	(f)	tan 30°	(i)	tan 45°	(l)	tan 60°

5.2 A Chain Device

The methods used in Section 5.1 can be extended to prove the irrationality of the trigonometric functions of any angle which is a whole number of degrees, minutes, and seconds, with but a few obvious exceptions. Thus we are speaking of angles like $14°41'33''$. Exceptions must be made for the angles $0°$, $30°$, $45°$, $60°$, and also any angles obtained from these four by adding or subtracting any integral multiple of $90°$. This is not to say that all the trigonometric functions of $30°$, for example, are rational, but at least one trigonometric function of $30°$ is rational.

These assertions will not be established in their broadest generality, because the equations arising for such numbers as $\cos 14°41'13''$ are too complex to be brought within our purview. Nevertheless there is one simple principle that will carry us a long way, namely:

If θ is any angle such that $\cos 2\theta$ *is irrational, then* $\cos\theta$, $\sin\theta$, *and* $\tan\theta$ *are also irrational.*

To prove this, we first use eq. (8). Suppose that $\cos\theta$ were rational. Then $\cos^2\theta$ and $2\cos^2\theta - 1$ also would be rational. But $2\cos^2\theta - 1$ is $\cos 2\theta$, which is irrational.

Similarly, suppose that $\sin\theta$ were rational. Then $\sin^2\theta$ would be rational and so would $1 - 2\sin^2\theta$. But this is $\cos 2\theta$ again.

Finally, suppose that $\tan\theta$ were rational. Then $\tan^2\theta$ would be rational, and then we could use the well-known trigonometric identity

$$1 + \tan^2\theta = \sec^2\theta = \frac{1}{\cos^2\theta}$$

to see that $\cos^2\theta$ would be rational. But again it would follow from eq. (8) that $\cos 2\theta$ is rational, and so we have a contradiction. Hence $\tan\theta$ must be irrational.

By repeated application of the principle we have just proved, we can establish the irrationality of infinitely many trigonometric numbers.

For example, from the irrationality of cos 20°, we conclude that

cos 10°	sin 10°	tan 10°
cos 5°	sin 5°	tan 5°
cos 2°30′	sin 2°30′	tan 2°30′
cos 1°15′	sin 1°15′	tan 1°15′
cos 37′30″	sin 37′30″	tan 37′30″
⋮	⋮	⋮

are irrational.

Problem Set 16

1. Prove that the following numbers are irrational
   - (a) cos 15°,   sin 15°,   tan 15°,
   - (b) cos 7°30′,   sin 7°30′,   tan 7°30′,
   - (c) cos 22°30′,   sin 22°30′,   tan 22°30′,
   - *(d) cos 35°,   sin 35°,   tan 35°,
   - *(e) cos 25°,   sin 25°,   tan 25°.

2. Prove that 14°41′13″ equals some rational number multiplied by 90°, i.e., prove that 14°41′13″ is a rational multiple of 90°.

3. (a) Prove that if cos θ is rational, then cos 3θ is also rational.
   (b) Is this equivalent to proving that if cos 3θ is irrational, then cos θ is irrational?

4. Prove that if sin 3θ is irrational, then sin θ is irrational.

5.3 Irrational Values of Common Logarithms

All the logarithms discussed in this book will be taken to the base 10, so there will be no need to specify this base in each case. We recall that, given a positive real number y, its logarithm to base 10 is defined to be a number k such that $10^k = y$. Thus for any $y > 0$,

$$\log y = k$$

and

$$10^k = y$$

are equivalent statements. All the proofs will be based on the Fundamental Theorem of Arithmetic, proved in Appendix B, that any integer has a unique factorization into primes.

EXAMPLE 1. Prove that log 2 is irrational.

SOLUTION. Suppose, on the contrary, that $\log 2 = a/b$, where a and b are positive integers. It is reasonable to take a and b positive, because $\log 2$ is positive. This equation means that

$$2 = 10^{a/b}.$$

Raising both sides to power b, we get

$$2^b = 10^a = 2^a 5^a.$$

This is an equality between positive integers, so the Fundamental Theorem of Arithmetic is applicable. In fact the Fundamental Theorem shows that this equation cannot hold because 2^b is an integer which is not divisible by 5 whatever value b may have, whereas $2^a 5^a$ is divisible by 5 since a is a positive integer. Hence $\log 2$ is irrational.

EXAMPLE 2. Prove that $\log 21$ is irrational.

SOLUTION. Suppose, on the contrary, that there are positive integers a and b such that

$$\log 21 = \frac{a}{b} \quad \text{or} \quad 21 = 10^{a/b}.$$

Again we raise both sides to power b to get

$$21^b = 10^a.$$

But this statement of equality cannot be true since 21^b has prime factors 3 and 7, whereas 10^a has prime factors 2 and 5.

EXAMPLE 3. Let c and d be two different non-negative integers. Prove that $\log (2^c 5^d)$ is irrational.

SOLUTION. Again we use an indirect argument. Because of the conditions imposed on c and d, we know that $2^c 5^d$ exceeds 1, so that $\log (2^c 5^d)$ is positive. Suppose that

$$\log (2^c 5^d) = \frac{a}{b},$$

where a and b are positive integers. Then we know that

$$2^c 5^d = 10^{a/b}.$$

Raising both sides to power b, we get

$$2^{bc} 5^{bd} = 10^a = 2^a 5^a.$$

According to the Fundamental Theorem of Arithmetic, this equation holds only if $bc = a$ and $bd = a$, i.e., if $bc = bd$. But since c and d are different integers, so also are bc and bd. Thus $\log (2^c 5^d)$ is irrational.

Problem Set 17

1. Prove that $\log 3/2$ is irrational.

2. Prove that $\log 15$ is irrational.

3. Prove that $\log 5 + \log 3$ is irrational.

*4. Prove that the integers $1, 2, 3, \ldots , 1000$ can be divided into three distinct non-overlapping classes.

Class A: the integers 1, 10, 100, 1000;
Class B: integers of the form $2^c 5^d$, where c and d are unequal;
Class C: integers divisible by at least one odd prime p, with p not equal
      to 5;
and that $\log n$ is rational if and only if n is in Class A.

5.4 Transcendental Numbers

Besides the classification of real numbers into rational and irrational, there is another separation into algebraic and transcendental. If a real number satisfies some equation of the the form

$$c_n x^n + c_{n-1} x^{n-1} + c_{n-2} x^{n-2} + \cdots + c_2 x^2 + c_1 x + c_0 = 0$$

with integral coefficients, we say that it is an *algebraic number*. If a real number satisfies no such equation, it is called a *transcendental number*. (Complex numbers are divided into algebraic and transcendental numbers in exactly the same way, but we will confine our attention to real numbers.)

It is easy to see that every rational number is an algebraic number. For example, 5/7 satisfies the equation $7x - 5 = 0$, and this equation is of the prescribed type. More generally, any rational number a/b satisfies the equation $bx - a = 0$, and so it is an algebraic number.

Since every rational number is algebraic, it follows that every non-algebraic number is non-rational (see [12] of "Twelve Ways of Stating 'If A then B' ", p. 27), or, stated more conventionally, *every transcendental number is irrational*. We can illustrate this schematically as in Fig. 15.

In the figure, we have listed $\sqrt{2}$ and $\sqrt[3]{7}$ as examples of algebraic numbers. They are algebraic because they satisfy the algebraic equations

$$x^2 - 2 = 0 \quad \text{and} \quad x^3 - 7 = 0$$

respectively. On the other hand, the numbers $\log 2$ and π have been listed as transcendental numbers. (The number π, with value $3.14159\cdots$ is the ratio of the length of the circumference to the length of the diameter of any circle.) We cannot prove here that these are transcendental numbers because such proofs involve much deeper methods than we are using. The transcendence of π has been known since 1882, but the transcendence of $2^{\sqrt{2}}$ and $\log 2$ are much more recent results, known only since 1934. The number $2^{\sqrt{2}}$ was used as a specific example by the great mathematician David Hilbert when, in 1900, he presented a famous list of twenty-three problems which he viewed as the outstand-

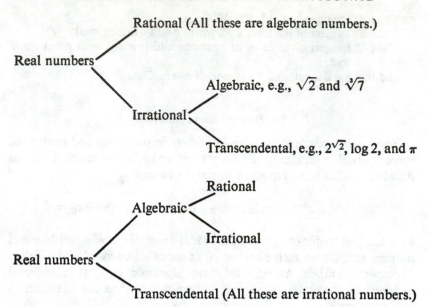

Figure 15

ing unsolved mathematical questions. Specifically, Hilbert's seventh problem was to decide whether α^β is algebraic or transcendental, given that α and β are algebraic numbers. (The cases $\alpha = 0$, $\alpha = 1$, and β rational were excluded because in these cases it is rather easy to prove that α^β is algebraic.) In 1934, it was settled by A. Gelfond and independently by Th. Schneider that α^β is transcendental. Of course, the transcendence of $2^{\sqrt{2}}$ is a special case of the general result. Another special consequence is the transcendence of log 2; for, if we denote log 2 by β, and 10 by α, then by the definition of common logarithm,

$$10^{\log 2} = \alpha^\beta = 2.$$

If β were algebraic and irrational, then by the Gelfond-Schneider Theorem, 2 would be transcendental. But this is not the case, so $\beta = \log 2$ is either rational or transcendental. We have seen that it is not rational. Hence log 2 is transcendental.

More generally, the Gelfond-Schneider theorem establishes the transcendence of log r, provided that r is rational and log r is irrational. In view of what we just proved in Section 5.3 (see also Problem 4 of Set 17), this says that log r is transcendental when r is any positive rational number except the following:

$$\ldots, 10^{-5}, 10^{-4}, 10^{-3}, 10^{-2}, 10^{-1}, 10^0, 10^1, 10^2, 10^3, 10^4, 10^5, \ldots.$$

It should be kept in mind that all logarithms mentioned in this book are common logarithms, that is logarithms to base 10.

Thus the numbers log n are *transcendental* if n is any integer between 1 and 1000 except $n = 1$, $n = 10$, $n = 100$, and $n = 1000$. On the other hand, the trigonometric numbers like cos 20° that were proved irrational in the early part of this chapter are *algebraic* numbers. The general result is this: let r be any rational number and let $(90r)°$ denote the angle obtained by multiplying 90° by r. Then

$$\sin (90r)°, \qquad \cos (90r)°, \qquad \text{and} \qquad \tan (90r)°$$

are algebraic numbers. (The only qualification that must be put on this statement is in the case of tan $(90r)°$, where the rational number r must be restricted to values for which this trigonometric function exists as a real number. For example, $r = 1$ is not admitted because tan 90° is not a real number.)

We said above that π is a transcendental number. This implies that π is an irrational number, and although it is easier to prove that π is irrational than that it is transcendental, even this is beyond the scope of this book.

Problem Set 18

1. Prove that the following numbers are algebraic:

   (a) $\sqrt{3}$,   (b) $\sqrt[3]{5}$,   (c) $\sqrt{2} + \sqrt{3}$,   (d) cos 20°,   (e) sin 10°

*2. Assuming that π is a transcendental number, prove that 2π is a transcendental number.

5.5 Three Famous Construction Problems

The theory of algebraic and transcendental numbers has enabled mathematicians to settle three well-known geometric problems that have come down from antiquity. These three problems, commonly referred to as "duplicating the cube," "trisecting an angle," and "squaring the circle," consist in performing the following constructions by means only of the straightedge and compass methods of Euclidean geometry:

(1) "Duplicating the cube," or "doubling the cube," means to construct a cube having twice the volume of a given cube. Although a cube is a figure in solid geometry, the problem is really one in plane geometry. For, if we take an edge of the given cube as the unit of length (Fig. 16), the problem is to construct a line of length $\sqrt[3]{2}$, because

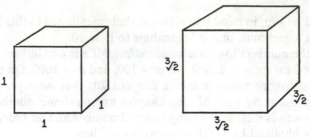

Figure 16

this would be the length of the edge of a cube that has twice the volume of the given cube.

(2) "Trisecting an angle" means to devise a method, using only the prescribed tools, by which *any* angle can be trisected. There are certain special angles, 45° and 90° for example, which can be trisected with straightedge and compass; but the so-called "general" angle cannot be divided into three equal angles with the prescribed tools.

(3) "Squaring the circle" means to construct a square equal in area to a given circle or, equivalently, to construct a circle equal in area to a given square.

These three constructions are now known to be impossible; that is, they cannot be done by the prescribed straightedge and compass methods of Euclidean geometry. Many amateurs continue to work on these problems, not knowing that their efforts are in vain. Although these amateurs are aware that no mathematician has yet been able to achieve these constructions, they are apparently not aware that the constructions have been proved impossible. What many an amateur mathematician achieves from time to time is an *approximate* solution to one of these problems, but never an exact solution. The distinction here is clear: the duplication of the cube problem, for example, is to produce a construction that would give, with theoretically perfect drawing instruments, a line not almost of length $\sqrt[3]{2}$, but exactly of length $\sqrt[3]{2}$. The problem is not solved, for instance, by a construction giving a line of length $10(8 - \sqrt{62})$, even though the numbers $10(8 - \sqrt{62})$ and $\sqrt[3]{2}$ agree to six decimal places.

A special source of misunderstanding occurs in the case of the angle trisection problem. It *is* possible to trisect any angle if markings are allowed on the straightedge. Thus the assertion of the impossibility of the trisection of a general angle can only be made with the understanding that the allowable construction processes involve the compass and the *unmarked* straightedge.

Because of the considerable confusion surrounding these three classic problems, we now give a rough notion of how it can be established that the constructions are impossible. Because the details become rather technical, we cannot give complete proofs. Nevertheless we hope to make the matter plausible. If any reader wants to tackle it, there is a full treatment of the trisection of the angle and the doubling of the cube in R. Courant and H. Robbins, *What Is Mathematics?*, Oxford University Press, pp. 127–138. The proof of the impossibility of squaring the circle is much more difficult than the other two proofs.

How is it possible to prove that these constructions are impossible? The first step is to get some idea of what kinds of lengths can be constructed with straightedge and compass, given a unit length. We state without proof (anyone familiar with geometric constructions will realize that the assertion is reasonable) that among the lengths that can be constructed are successions of square roots applied to rational numbers, for instance,

(10) $\quad \sqrt{2}, \ \sqrt{1+\sqrt{2}}, \ \sqrt{5-3\sqrt{1+\sqrt{2}}}, \ \sqrt{1+\sqrt{5-3\sqrt{1+\sqrt{2}}}}$

These numbers are all algebraic. The four that are listed as examples in (10) are roots respectively of the equations

(11) $\qquad\qquad\qquad x^2 - 2 = 0,$

(12) $\qquad\qquad\qquad x^4 - 2x^2 - 1 = 0,$

(13) $\qquad\qquad x^8 - 20x^6 + 132x^4 - 320x^2 + 94 = 0,$

(14) $\qquad\quad x^{16} - 8x^{14} + 8x^{12} + 64x^{10} - 98x^8$
$\qquad\qquad\quad - 184x^6 + 200x^4 + 224x^2 - 113 = 0.$

Let us choose one of these, say (13), and verify it. We begin with

$$x = \sqrt{5 - 3\sqrt{1 + \sqrt{2}}}.$$

Squaring, we get

$$x^2 = 5 - 3\sqrt{1 + \sqrt{2}}.$$

Moving a term and squaring again, we obtain

$$x^2 - 5 = -3\sqrt{1 + \sqrt{2}},$$
$$x^4 - 10x^2 + 25 = 9 + 9\sqrt{2},$$
$$x^4 - 10x^2 + 16 = 9\sqrt{2}.$$

A final squaring of both sides leads to (13).

Now, not only are the numbers (10) roots of eqs. (11) to (14), but none of these numbers is a root of an equation of lower degree with integer coefficients. Take the number $\sqrt{1 + \sqrt{2}}$, for example. It satisfies eq. (12) of degree 4, but it satisfies no equation of degree 3, 2, or 1 with integer coefficients. (We do not prove this assertion.) Whenever an algebraic number is a root of an equation of degree n with integer coefficients, but is a root of no equation of lower degree with integer coefficients, we say that it is *an algebraic number of degree n.* Thus the numbers (10) are algebraic numbers of degree 2, 4, 8, and 16 respectively. This suggests the following basic truth about lengths that can be constructed by the methods of Euclidean geometry:

THEOREM ON GEOMETRIC CONSTRUCTIONS. *Beginning with a line segment of unit length, any length that can be constructed by straightedge and compass methods is an algebraic number of degree 1, or 2, or 4, or 8, . . . , i.e., in general, an algebraic number whose degree is a power of 2.*

If the reader will take this result for granted, we can indicate how it happens that the three famous constructions are impossible.†

Let us begin with the duplication, or doubling, of the cube. As we saw when we stated the problem, it amounts to constructing a line of length $\sqrt[3]{2}$ from a given unit length. But is $\sqrt[3]{2}$ a constructible length? It satisfies the equation

$$(15) \qquad\qquad x^3 - 2 = 0,$$

and this suggests that $\sqrt[3]{2}$ is an algebraic number of degree 3. Indeed this is so, and in order to prove it we must establish only that $\sqrt[3]{2}$ satisfies no equation of degree 1 or degree 2 with integer coefficients. Although this is not difficult, it is a little tricky, and we shall postpone this proof until the next section.

Since $\sqrt[3]{2}$ is an algebraic number of degree 3, by the Theorem on Geometric Constructions, stated above, it is not constructible. Hence we conclude that it is impossible to duplicate the cube.

Consider next the problem of the trisection of the angle. To establish that this is impossible, it is enough to show that a specific angle cannot be trisected by the prescribed methods. The specific angle that we take is 60°. To trisect an angle of 60° means the construction of a 20° angle.

---

† The reader will recall (see p. 27, [12]) that this theorem implies the statement: Algebraic numbers of degree m, where m is *not* a power of 2, are *not* constructible by straightedge and compass; also transcendental numbers are not so constructible.

This amounts to constructing, from a given line segment of length 1, a line segment whose length is cos 20°. To see this, consider a triangle of base 1 with base angles 60° and 90°, as shown in Fig. 17. Thus we

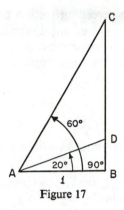

Figure 17

have a triangle *ABC* with base *AB* = 1, angle *BAC* = 60°, angle *ABC* = 90°. On the line *BC*, let the point *D* be chosen in such a way that angle *BAD* = 20°. From elementary trigonometry, we know that

$$AD = \frac{AD}{1} = \frac{AD}{AB} = \sec 20°.$$

Thus the trisection of the 60° angle amounts to the construction of a line segment of length sec 20°. But this in turn amounts to the construction of a line segment of length cos 20°, because cos 20° and sec 20° are reciprocals, and it is well known that if a certain segment is constructible, then the segment of reciprocal length is also constructible.

So the question is: Can a line segment of length cos 20° be constructed from a given line segment of length 1? We know from eq. (6) that cos 20° is a root of a cubic equation, i.e., an equation of degree 3. Moreover we state (without proof, for it is a little deep) that cos 20° satisfies no equation of degree 1 or degree 2 with integer coefficients. Thus cos 20°, like $\sqrt[3]{2}$, is an algebraic number of degree 3, and so by the Theorem on Geometric Constructions, cos 20° is not constructible. Thus the trisection of the 60° angle is impossible by straightedge and compass methods.

Finally, consider the problem of squaring the circle. Given any circle, we may consider its radius as the unit of length. With this unit, the area of the circle is π square units. A square of equal size would have a side length of $\sqrt{\pi}$. So the problem of squaring the circle amounts to constructing a line of length $\sqrt{\pi}$ from a given unit length. Now it is well known in geometric construction theory that from line segments of lengths 1 and a, a line segment of length a^2 can be constructed. So if a line segment of length $\sqrt{\pi}$ could be constructed, so also could a line segment of length π.

But we asserted in the preceding section that π is a transcendental number, that is to say, π is not an algebraic number. Hence by the Theorem on Geometric Constructions, it is not possible to construct a line segment of length π. Thus the "squaring the circle" construction is impossible.

Problem Set 19

(Problems 2 and 3 are for students with a knowledge of geometric constructions.)

1. Prove that the first, second, and fourth numbers in the list (10) are roots of equations (11), (12), and (14) respectively.

2. Prove that, given line segments of length 1 and length sin 20°, a line segment of length cos 20° can be constructed by the straightedge and compass methods.

3. Prove that, given line segments of length 1 and length tan 20°, a line segment of length cos 20° can be constructed by the straightedge and compass methods.

5.6 Further Analysis of $\sqrt[3]{2}$

In the preceding section, we asserted that $\sqrt[3]{2}$ is an algebraic number of degree 3. That is, $\sqrt[3]{2}$ is a root of the equation $x^3 - 2 = 0$, but it is a root of no equation of degree 1 or degree 2 with integer coefficients. We now prove this assertion.

To establish that $\sqrt[3]{2}$ is a root of no equation of degree 1 with integer coefficients, we must prove that there are no integers a and b, with a not equal to zero, so that

$$a \sqrt[3]{2} + b = 0.$$

If there were any such integers, then we would have $\sqrt[3]{2} = -b/a$, so that $\sqrt[3]{2}$ would be a rational number. But we know that $\sqrt[3]{2}$ is irrational, by Corollary 2 of Section 4.3.

It is more difficult to prove that $\sqrt[3]{2}$ is not a root of any quadratic equation with integer coefficients, such as

$$ax^2 + bx + c = 0.$$

We presume that $\sqrt[3]{2}$ is a root of such an equation, and then we deduce a contradiction. Thus we presume that

$$a(\sqrt[3]{2})^2 + b\sqrt[3]{2} + c = 0,$$

i.e., that

$$a\sqrt[3]{4} + b\sqrt[3]{2} = -c.$$

Squaring both sides and simplifying, we obtain

$$b^2\sqrt[3]{4} + 2a^2\sqrt[3]{2} = c^2 - 4ab.$$

The last two equations can be thought of as two simultaneous linear equations in the quantities $\sqrt[3]{4}$ and $\sqrt[3]{2}$. Either they can be solved or they cannot, depending on whether the pairs of coefficients, a,b and $b^2, 2a^2$, are not, or are, proportional.

If they can be solved, for example by eliminating $\sqrt[3]{4}$, we get

$$\sqrt[3]{2} = \frac{4a^2b - ac^2 - b^2c}{b^3 - 2a^3}.$$

But $\sqrt[3]{2}$ is irrational, so we have a contradiction.

The other possibility is that the pairs of coefficients are proportional. This means that

$$\frac{a}{b} = \frac{b^2}{2a^2}, \qquad 2 = \frac{b^3}{a^3} = \left(\frac{b}{a}\right)^3, \qquad \sqrt[3]{2} = \frac{b}{a}.$$

Again we have a contradiction, and so we conclude that $\sqrt[3]{2}$ is an algebraic number of degree 3.

Problem Set 20

1. Prove that $\sqrt{2}$ is an algebraic number of degree 2.
2. Prove that $\sqrt[3]{3}$ is an algebraic number of degree 3.

5.7 A Summary

In this chapter we have applied the methods developed earlier to show that trigonometric functions and common logarithms have irrational values in most of the cases listed in elementary tables. Then we divided the real numbers into two new classes, the algebraic and the transcendental numbers, and saw how these classes were related to the earlier separation of real numbers into rational and irrational numbers. Without proof, we learned that if a length can be constructed from a given unit length by straightedge and compass, then that length is an algebraic number of degree 2^k, where k is some non-negative integer. (A reader who is quite familiar with analytic geometry can persuade himself somewhat of the truth of this theorem regarding geometric constructions by analyzing the algebraic meaning of the steps which can be performed with straightedge and compass. The three cases are a straight line intersecting a straight line, a straight line intersecting a circle, and a circle intersecting a circle.) Having thus eliminated the possibility of constructing a straight line whose length is an algebraic number of degree 3, we have seen that a cube cannot be doubled, nor a general angle trisected, by the prescribed methods. Because of the impossibility of constructing any straight line whose length is a transcendental number, we have seen how the problem of squaring the circle was settled: this construction too is not possible.

CHAPTER SIX

The Approximation of Irrationals by Rationals

This chapter is concerned with the closeness of approximation of an irrational number by rational numbers. Now, as we shall see, we can get rational numbers that are as close to $\sqrt{2}$, for example, as we please. There are rational numbers a/b which are within 10^{-10} of $\sqrt{2}$, or within 10^{-20} of $\sqrt{2}$, or within any other range of difference that we want to specify. And this is true for any irrational number, not only for $\sqrt{2}$.

But in order to find a rational number a/b which differs from an irrational number by less than 10^{-20}, we must look for an a/b with a very large denominator b. If we allow b to be as large as 10^{20}, we can find a fraction a/b to meet the specifications. What happens if we restrict b to be no larger than 10^{15} or 10^{10}? The problem becomes deeper and more troublesome. In looking at questions of this kind, we will be concerned with what can be said for *every* irrational number, not just for some special cases like $\sqrt{2}$ and $\sqrt{3}$.

To talk about the approximation of one number by another, we must have available the language and notation of inequalities. Therefore, we shall begin with this topic.

6.1 Inequalities†

We write $u > v$ whenever u is greater than v, and this means that $u - v$ is positive. Of course, whenever u is greater than v, it follows

† For a thorough treatment of inequalities, see E. Beckenbach and R. Bellman, *An Introduction to Inequalities*, in this series.

that v is less than u, written as $v < u$ in mathematical notation. Consequently the four inequalities

$$u > v, \qquad u - v > 0, \qquad v < u, \qquad v - u < 0$$

are simply four ways of stating the same basic relationship between u and v. Similarly, $u \geqq v$ means that u is greater than or equal to v, and this amounts to saying that $u - v$ is positive or zero, but not negative.

THEOREM 6.1.

(a) *If $u > v$, and w is any number, then $u + w > v + w$.*
(b) *If $u > v$, and w is any number, then $u - w > v - w$.*
(c) *If $u > v$, and w is any positive number, then $uw > vw$.*
(d) *If $u > v$, and w is any positive number, then $u/w > v/w$.*
(e) *If $u > v$, and if u and v are positive, then $u^2 > v^2$ but $1/u < 1/v$.*
(f) *If $u > v$, and $v > w$, then $u > w$.*
(g) *All these results hold if the inequality signs, $>$ and $<$, are replaced by $\geqq$ and $\leqq$ throughout.*

PROOF: We shall take two principles for granted: that the sum and the product of two positive numbers are positive.

(a) We are given that $u - v$ is positive, and we are to prove that $(u + w) - (v + w)$ is positive. This is clear, because

$$u - v = (u + w) - (v + w).$$

(b) Again we are given that $u - v$ is positive, and we are to prove that $(u - w) - (v - w)$ is positive. As before, this follows because

$$u - v = (u - w) - (v - w).$$

(c) We are given that $u - v$ and w are positive, and we are to prove that $uw - vw$ is positive. This follows from the facts that $uw - vw = w(u - v)$, and that the product of two positive numbers is positive.

(d) This is really contained in part (c), because if w is positive, so is $1/w$. Consequently, $1/w$ could be used as the multiplier in part (c) in place of w, thus: if $u > v$, then $u(1/w) > v(1/w)$.

(e) Since u and v are positive, so is $u + v$. But $u > v$ implies that $u - v$ is also positive, and hence that the product $(u + v)(u - v)$ is positive. Thus we have

$$(u + v)(u - v) > 0, \qquad u^2 - v^2 > 0, \qquad u^2 > v^2.$$

On the other hand, if we use part (c) to justify the multiplication of both sides of $u > v$ by $1/uv$, we get

$$u \cdot \frac{1}{uv} > v \cdot \frac{1}{uv}$$

and hence

$$\frac{1}{v} > \frac{1}{u} \quad \text{or} \quad \frac{1}{u} < \frac{1}{v}.$$

(f) We are given that $u - v$ and $v - w$ are positive, and we want to prove that $u - w$ is positive. However,

$$u - w = (u - v) + (v - w),$$

so once again we have merely to use the principle that the sum of two positive numbers is positive.

(g) One way to prove part (g) would be to run through parts (a), (b), ..., (f), giving a fresh analysis in each case. However, there is an easier way. The proofs of parts (a) to (f) have been based on the principles that the sum and the product of two positive numbers are themselves positive. Now, whereas $u > v$ means that $u - v$ is positive, $u \geqq v$ means that $u - v$ is positive or zero, that is to say, $u - v$ is non-negative. Furthermore, the sum and the product of any two non-negative numbers are themselves non-negative, and on this basis it must be that all the proofs (a) to (f) will automatically extend from the $>$ case to the $\geqq$ case.

If the numbers u and v are plotted on the real line, as explained in Chapter 3, the inequality $v < u$ has the significance that v is to the left of u, or u is to the right of v (Fig. 18). The inequalities $w < v < u$ are

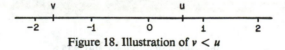

Figure 18. Illustration of $v < u$

taken to mean that $w < v$ and $v < u$, so that v is between w and u (Fig. 19). We must explain the use of the word "between."

Figure 19. Illustration of $w < v < u$

If we write $w < v < u$, we mean v is "strictly between" w and u and may not coincide with either w or u; but if we say "between" or write $w \leqq v \leqq u$, we include the possibilities of v being equal to w or to u. We may want to admit only one of these possibilities, for instance $w < v \leqq u$ or $w \leqq v < u$. The symbols make the meaning perfectly clear.

Problem Set 21

1. Given that $u^2 > v^2$ and that u and v are positive, prove that $u > v$.

2. Given that $r > s$, prove that $-r < -s$.

3. Prove that terms in inequalities can be transposed from one side to the other provided the signs are changed. Specifically, prove that if $a + b - c > d + e - f$, then $a - e + f > d - b + c$.

4. Given positive integers n and k such that $n \leqq k$, prove that

$$\frac{1}{n} \geqq \frac{1}{k} \quad \text{and} \quad \frac{1}{n^2} \geqq \frac{1}{kn}.$$

5. Determine the truth or falsity of the following:
   (a) If $r > s$, then $r^2 > s^2$.
   (b) If $r > s$, and c is any number, then $cr > cs$.
   (c) If $-1/2 < \lambda < 1/2$ holds, then $-1 < \lambda < 1$ holds.†
   (d) If $-1/2 < \lambda < 1/2$ holds, then $-3/2 < \lambda < 3/2$ holds.
   (e) If $0 < \lambda < 1/2$ holds, then $-1/2 < \lambda < 1/2$ holds.
   (f) If $-1/2 < \lambda < 1/2$ holds, then $-1/3 < \lambda < 1/3$ holds.
   (g) If $-1/2n < \lambda < 1/2n$ holds, then $-1/n < \lambda < 1/n$ holds.

6. A certain irrational number λ lies strictly between -10 and 10. Write this in mathematical notation.

7. If w is negative and $u > v$, prove that $uw < vw$.

8. (a) Let u and v be any two different integers chosen from $1, 2, 3, \ldots , 10$. Prove that $-9 \leqq u - v \leqq 9$.
   (b) If in (a) we did not prescribe that the integers u and v be different, would it still be true that $-9 \leqq u - v \leqq 9$?

6.2 Approximation by Integers

If we round off any real number by replacing it with the closest integer, the error committed will be at most $1/2$. For example, if we replace 6.3 by 6, or 9.7 by 10, or 7.5 by either 7 or 8, the error is no more than $1/2$ in each case. If we replace an irrational number by the nearest integer, the error is less than $1/2$, and we begin the theory of approximations with this simple case.

† The symbol λ is a Greek letter, read "lambda."

THEOREM 6.2. *Corresponding to any irrational number α there is a unique integer m such that*

$$-\frac{1}{2} < \alpha - m < \frac{1}{2}.$$

PROOF. We choose m as the integer closest to α. For example, if $\alpha = \sqrt{3} = 1.73 \cdots$, we choose $m = 2$, or if $\alpha = 2\sqrt{3} = 3.46 \cdots$, we choose $m = 3$. Thus m may be the integer just larger than α, or just smaller than α, whichever is closer. (It is clear that one of them is closer to α than the other, otherwise α would be halfway between two consecutive integers, say n and $n + 1$. But then α would equal $n + 1/2$ which is a rational number, and this contradicts our assumption.) We can state this in another way. Any line segment AB of unit length (i.e., one unit long) marked off on the real line, as in Fig. 20, will contain

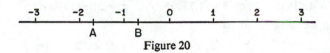

Figure 20

exactly one integer unless A and B happen to be integer points. Now take A to be the point corresponding to the number $\alpha - 1/2$ and B corresponding to $\alpha + 1/2$. Since $\alpha - 1/2$ and $\alpha + 1/2$ are not integers (they are not even rational—see Theorem 4.1, Chapter 4), we know that A and B cannot be integer points. Denoting the unique integer in the segment AB by m, we see that m lies strictly between $\alpha - 1/2$ and $\alpha + 1/2$. Thus

$$\alpha - \frac{1}{2} < m < \alpha + \frac{1}{2}.$$

Subtracting α, we get

$$-\frac{1}{2} < m - \alpha < \frac{1}{2}.$$

Now if a number $m - \alpha$ lies between $-1/2$ and $1/2$, so will the number obtained by changing the sign, and hence $\alpha - m$ also lies between $-1/2$ and $1/2$. Thus we get the inequalities of Theorem 6.2.

The integer m is unique, because if there were another integer n satisfying

$$-\frac{1}{2} < \alpha - n < \frac{1}{2},$$

then n would also satisfy

$$-\frac{1}{2} < n - \alpha < \frac{1}{2}.$$

Adding α to these inequalities, we see that n would satisfy

$$\alpha - \frac{1}{2} < n < \alpha + \frac{1}{2}.$$

But the segment AB contained only one integer, so n is the same as m.

Problem Set 22

(In this and subsequent problem sets, the reader may wish to use $\sqrt{2} = 1.41421\cdots$, $\sqrt{3} = 1.73205\cdots$, $\pi = 3.14159\cdots$.)

1. Find the integers closest to

    (a) $\sqrt{2}$,   (b) $2\sqrt{2}$,   (c) $3\sqrt{2}$,   (d) $4\sqrt{2}$,   (e) $3\sqrt{3}$

    (f) $4\sqrt{3}$,   (g) π,   (h) 10π,   (i) $-\sqrt{3}$,   (j) -7π

2. Given any irrational number α, prove that there is a unique integer q such that $0 < \alpha - q < 1$.

6.3 Approximation by Rationals

One way to approximate an irrational number, such as $\sqrt{2}$, is to use the decimal form

$$\sqrt{2} = 1.41421\cdots.$$

The numbers 1, 1.4, 1.41, 1.414, 1.4142, 1.41421, ... form a series of closer and closer approximations to $\sqrt{2}$. These approximations are all rational numbers, and so we have an infinite sequence of rational approximations to $\sqrt{2}$:

(1) $\dfrac{1}{1}, \dfrac{14}{10}, \dfrac{141}{100}, \dfrac{1414}{1000}, \dfrac{14142}{10000}, \dfrac{141421}{100000}, \cdots.$

These numbers are getting closer and closer to $\sqrt{2}$ as we go further along in the series. Furthermore, we can write the inequalities

$$\frac{1}{1} < \sqrt{2} < \frac{2}{1},$$

$$\frac{14}{10} < \sqrt{2} < \frac{15}{10},$$

$$\frac{141}{100} < \sqrt{2} < \frac{142}{100},$$

$$\frac{1414}{1000} < \sqrt{2} < \frac{1415}{1000},$$

$$\frac{14142}{10000} < \sqrt{2} < \frac{14143}{10000},$$

$$\frac{141421}{100000} < \sqrt{2} < \frac{141422}{100000}, \text{ etc.}$$

These inequalities show that infinitely many of the terms of (1) are as close to $\sqrt{2}$ as we want to specify. Suppose, for example, we want to know that there are infinitely many rational numbers within 0.0001 of $\sqrt{2}$. These we can get by taking all but the first four terms of the sequence (1).

However, the rational numbers (1) have the special feature that their denominators are powers of 10. There might be better approximations to $\sqrt{2}$ among rational numbers in general, without any restriction on their denominators.

Let us turn to the irrational number π for a well-known instance that illustrates our discussion. Since π has the value $3.14159 \cdots$, the sequence for π analogous to (1) is

(2) $\qquad \dfrac{3}{1}, \ \dfrac{31}{10}, \ \dfrac{314}{100}, \ \dfrac{3141}{1000}, \ \dfrac{31415}{10000}, \ \dfrac{314159}{100000}, \ \cdots.$

However, we know that 22/7 is a closer approximation to π than 31/10. In fact 22/7 is closer than 314/100, although not closer than subsequent terms of the sequence (2).

To get away from our dependence on the denominators 10, 10^2, 10^3, etc., we first show that every irrational number can be approximated by a rational number having any given denominator.

THEOREM 6.3. *Let λ be any irrational number and n be any positive integer. Then there is a rational number with denominator n, say m/n, such that*

$$-\frac{1}{2n} < \lambda - \frac{m}{n} < \frac{1}{2n}.$$

We motivate the proof of this theorem with an example. Suppose that λ is $\sqrt{2}$ and n is 23. Consider the irrational number $23\sqrt{2}$, which, by use of the decimal expansion $1.41421\cdots$ for $\sqrt{2}$, has the approximate value

$$23\sqrt{2} = 32.52\cdots$$

Thus the closest integer to $23\sqrt{2}$ is 33, and this is the "m" of Theorem 6.2 which, for $\alpha = 23\sqrt{2}$, states that

$$-\frac{1}{2} < 23\sqrt{2} - 33 < \frac{1}{2}.$$

But 33 is also the "m" of Theorem 6.3: for, according to Theorem 6.1, we can divide these inequalities by 23 to obtain

$$-\frac{1}{46} < \sqrt{2} - \frac{33}{23} < \frac{1}{46}.$$

PROOF. In general, beginning with any irrational λ and any positive integer n, we note that by Theorem 4.1 of Chapter 4, $n\lambda$ is irrational. Then we define m as the nearest integer to $n\lambda$, and so by Theorem 6.2,

$$-\frac{1}{2} < n\lambda - m < \frac{1}{2}.$$

By Theorem 6.1, these inequalities can be divided by the positive integer n to give

$$-\frac{1}{2n} < \lambda - \frac{m}{n} < \frac{1}{2n}.$$

Hence Theorem 6.3 is proved.

EXAMPLE. Find rational numbers m/n as in Theorem 6.3 for the case where $\lambda = \sqrt{2}$, for $n = 1, 2, 3, 4, 5, 6, 7, 8, 9, 10$.

SOLUTION. By simple calculation the integers nearest to

$$\sqrt{2}, \quad 2\sqrt{2}, \quad 3\sqrt{2}, \quad 4\sqrt{2}, \quad 5\sqrt{2}, \quad 6\sqrt{2}, \quad 7\sqrt{2}, \quad 8\sqrt{2}, \quad 9\sqrt{2}, \quad 10\sqrt{2}$$

are $1, 3, 4, 6, 7, 8, 10, 11, 13, 14$. Hence the rational numbers to be determined are

$$\frac{1}{1}, \quad \frac{3}{2}, \quad \frac{4}{3}, \quad \frac{6}{4}, \quad \frac{7}{5}, \quad \frac{8}{6}, \quad \frac{10}{7}, \quad \frac{11}{8}, \quad \frac{13}{9}, \quad \frac{14}{10},$$

and the error in each of these approximations is less than $1/2n$, where n is the integer in the denominator.

This example shows that the rational fractions m/n of Theorem 6.3 are not necessarily in lowest terms.

Problem Set 23

1. Find rational numbers m/n as in Theorem 6.3 for the case where $\lambda = \sqrt{3}$ and $n = 1, 2, 3, 4, 5, 6, 7, 8, 9, 10$.

2. Find rational numbers m/n as in Theorem 6.3 for the case where $\lambda = \pi = 3.14159\cdots$, for $n = 1, 2, 3, 4, 5, 6, 7, 8, 9, 10$.

3. Given any irrational number λ and any positive integer n, prove that there is an integer m such that

$$-\frac{1}{n} < \lambda - \frac{m}{n} < \frac{1}{n}.$$

*4. For a fixed irrational number λ and a fixed positive integer n, prove that there is only one integer m that satisfies the inequalities of Theorem 6.3.

*5. Prove that Theorem 6.3 would be false if the words "in lowest terms" were inserted with reference to m/n, thus: "... Then there is a rational number in lowest terms with denominator n, say m/n, ..."

6.4 Better Approximations

Theorem 6.3 states that any irrational number λ can be approximated by a rational number m/n "to within $1/2n$," i.e., with an error less than $1/2n$. Can this be done to within $1/3n$, or $1/4n$, or perhaps closer? The answer is yes. In the next theorem we show that λ can be approximated by m/n to within $1/kn$ for any k we wish to specify: $k = 3$, $k = 4$, $k = 1000$, etc. But, whereas in Theorem 6.3 the approximation to within $1/2n$ could be achieved for every positive integer n, the approximation to within $1/kn$ with prescribed k in Theorem 6.4 cannot be obtained for all n.

Can we approximate any irrational number λ by m/n to within $1/n^2$, or $1/n^3$, or even closer? To within $1/n^2$, yes; to within $1/n^3$, no. But these are the topics of later sections. So let us begin with approximations of λ by m/n to within $1/kn$.

THEOREM 6.4. *Given any irrational number λ and any positive integer k, there is a rational number m/n whose denominator n does not exceed k, such that*

$$-\frac{1}{nk} < \lambda - \frac{m}{n} < \frac{1}{nk}.$$

Before presenting a proof of Theorem 6.4 that is valid for any λ and k, we shall prove the theorem for a particular instance, namely with $\lambda = \sqrt{3}$ and $k = 8$. First we enumerate the multiples of λ from $1 \cdot \lambda$ to $k \cdot \lambda$. We list the multiples of $\sqrt{3}$, writing each multiple as a sum of two positive numbers, an integer and a number less than one:

$$\sqrt{3} = 1 + 0.732\cdots, \qquad \sqrt{3} - 1 = 0.732\cdots,$$
$$2\sqrt{3} = 3 + 0.464\cdots, \qquad 2\sqrt{3} - 3 = 0.464\cdots,$$
$$3\sqrt{3} = 5 + 0.196\cdots, \qquad 3\sqrt{3} - 5 = 0.196\cdots,$$
$$4\sqrt{3} = 6 + 0.928\cdots, \qquad 4\sqrt{3} - 6 = 0.928\cdots,$$
$$5\sqrt{3} = 8 + 0.660\cdots, \qquad 5\sqrt{3} - 8 = 0.660\cdots,$$
$$6\sqrt{3} = 10 + 0.392\cdots, \qquad 6\sqrt{3} - 10 = 0.392\cdots,$$
$$7\sqrt{3} = 12 + 0.124\cdots, \qquad 7\sqrt{3} - 12 = 0.124\cdots,$$
$$8\sqrt{3} = 13 + 0.856\cdots, \qquad 8\sqrt{3} - 13 = 0.856\cdots.$$

The expressions in the right-hand column were obtained from those on the left by subtracting the integer part.

Next, we separate the unit interval into eight parts, $I_1, I_2, \ldots, I_8$, as

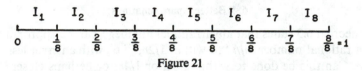

Figure 21

shown in Fig. 21. Thus I_1 consists of the numbers between 0 and 1/8, I_2 the numbers between 1/8 and 2/8, I_3 the numbers between 2/8 and 3/8, and so on.† Then we classify the eight decimal parts of the multiples of $\sqrt{3}$ into the categories $I_1, I_2, \ldots, I_8$ as follows:

$0.732\cdots$ is in I_6 (because $0.732\cdots$ is between 5/8 and 6/8),
$0.464\cdots$ is in I_4,
$0.196\cdots$ is in I_2,
$0.928\cdots$ is in I_8,
$0.660\cdots$ is in I_6,
$0.392\cdots$ is in I_4,
$0.124\cdots$ is in I_1,
$0.856\cdots$ is in I_7.

† Since we wish to derive strict inequalities, it is convenient to interpret "between" as "strictly between"; thus the intervals I_j include all points u such that

$$(j-1)/8 < u < j/8.$$

We use the number in the above list which is in I_1:

$$0.124 \cdots \text{ is in } I_1; \text{ that is, } 7\sqrt{3} - 12 \text{ is in } I_1.$$

But the numbers in I_1 lie between 0 and 1/8, so

$$0 < 7\sqrt{3} - 12 < \frac{1}{8}.$$

Now, since the number $7\sqrt{3} - 12$ lies between 0 and 1/8, it certainly lies between $-1/8$ and 1/8, thus

$$-\frac{1}{8} < 7\sqrt{3} - 12 < \frac{1}{8}.$$

Dividing this inequality by 7, we get

$$-\frac{1}{7 \cdot 8} < \sqrt{3} - \frac{12}{7} < \frac{1}{7 \cdot 8}.$$

This is a result of the form stated in Theorem 6.4, with $k = 8$, $n = 7$, and $m = 12$.

Our argument was based on the fact that $7\sqrt{3} - 12$ was in I_1. What would we have done if there had been no number in the interval I_1? The answer is that if there had been no number in the interval I_1, then one of the intervals $I_2, I_3, \ldots, I_7$ would contain two or more numbers. In the present example, not only is there a number in I_1, but also there are two in I_4 and two in I_6. Consider the pair in I_6:

$$0.732 \cdots \text{ is in } I_6; \text{ that is, } \sqrt{3} - 1 \text{ is in } I_6;$$

and

$$0.660 \cdots \text{ is in } I_6; \text{ that is, } 5\sqrt{3} - 8 \text{ is in } I_6.$$

Now, whenever two numbers are in I_6 (or in any one of these intervals), they are within 1/8 of each other, so their difference lies between $-1/8$ and $+1/8$. In particular, for the two numbers in I_6, we have

$$-\frac{1}{8} < (5\sqrt{3} - 8) - (\sqrt{3} - 1) < \frac{1}{8},$$

$$-\frac{1}{8} < 4\sqrt{3} - 7 < \frac{1}{8}.$$

Dividing by 4 we get

$$-\frac{1}{4\cdot 8} < \sqrt{3} - \frac{7}{4} < \frac{1}{4\cdot 8},$$

and this is another result of the form stated in Theorem 6.4 for $\lambda = \sqrt{3}$ and $k = 8$, this time with $n = 4$ and $m = 7$.

PROOF OF THEOREM 6.4. The special instances we have just discussed can serve as models for the proof of Theorem 6.4. Given an irrational number λ and a positive integer k, we take the k numbers λ, 2λ, 3λ, 4λ, ..., $k\lambda$ and write each of these numbers as an integer plus a fractional or decimal part:

$$\begin{aligned}
\lambda &= a_1 + \beta_1, & \lambda - a_1 &= \beta_1, \\
2\lambda &= a_2 + \beta_2, & 2\lambda - a_2 &= \beta_2, \\
3\lambda &= a_3 + \beta_3, & 3\lambda - a_3 &= \beta_3, \\
4\lambda &= a_4 + \beta_4, & 4\lambda - a_4 &= \beta_4, \\
&\ \ \vdots & &\ \ \vdots \\
k\lambda &= a_k + \beta_k, & k\lambda - a_k &= \beta_k.
\end{aligned}$$

The symbols a_1, a_2, ..., a_k stand for integers, whereas the symbols β_1, β_2, ..., β_k stand for numbers between 0 and 1. Next we divide the unit interval into k parts, I_1, I_2, ..., I_k, each of length $1/k$ (Fig. 22).

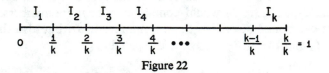

Figure 22

Thus the interval I_1 consists of the numbers between 0 and $1/k$, I_2 the numbers between $1/k$ and $2/k$, I_3 the numbers between $2/k$ and $3/k$, etc. The word "between" is here used in the strict sense, so that, for example, the numbers $2/k$ and $3/k$ are not themselves members of the interval I_3. Note that, by Theorem 4.1 of Chapter 4, each of the numbers β_1, β_2, ..., β_k is irrational. Consequently no β can equal any of the rational numbers

$$0, \ \frac{1}{k}, \ \frac{2}{k}, \ \frac{3}{k}, \ \ldots, \ \frac{k-1}{k}, \ \frac{k}{k}.$$

Thus each β lies in exactly one of the intervals I_1, I_2, I_3, ..., I_k.

There are two possibilities regarding I_1: either I_1 contains one or more of the β's, or I_1 contains none of the β's. We shall treat these two possibilities separately.

Case 1. The interval I_1 contains one or more β's. Thus there is one β, say β_n, in the interval I_1. The symbol n stands for some integer among $1, 2, 3, \ldots, k$. The number β_n is the same as $n\lambda - a_n$, so we know that

$$0 < n\lambda - a_n < \frac{1}{k}$$

because I_1 is the interval from 0 to $1/k$. It follows that

$$-\frac{1}{k} < n\lambda - a_n < \frac{1}{k},$$

and if we divide through by n we get

$$-\frac{1}{kn} < \lambda - \frac{a_n}{n} < \frac{1}{kn}.$$

Thus Theorem 6.4 is proved in this case, because we can define m to be the integer a_n.

Case 2. The interval I_1 contains none of the β's. In this case, the k numbers lie in the $k-1$ intervals

$$I_2, \quad I_3, \quad \ldots, \quad I_{k-1}.$$

At this point we apply the *Dirichlet pigeon-hole principle*, which says that if there are k pigeons in $k-1$ holes, there must be at least one pigeon-hole with two or more pigeons in it. Thus there must be at least one interval containing two or more β's. Let us say that β_r and β_j are in the same interval, where r and j are two different numbers among $1, 2, 3, \ldots, k$. Presuming that j is larger than r, we then know that $j - r$ is a positive integer less than k.

Since β_r and β_j lie inside the same interval of length $1/k$, their difference lies between $-1/k$ and $1/k$. Thus

$$-\frac{1}{k} < \beta_j - \beta_r < \frac{1}{k}.$$

But $\beta_j = j\lambda - a_j$ and $\beta_r = r\lambda - a_r$, so that

$$-\frac{1}{k} < (j\lambda - a_j) - (r\lambda - a_r) < \frac{1}{k}$$

or

$$-\frac{1}{k} < (j - r)\lambda - (a_j - a_r) < \frac{1}{k}.$$

Define n as $j - r$, and m as $a_j - a_r$, and then we have

$$-\frac{1}{k} < n\lambda - m < \frac{1}{k}.$$

We see that n is a positive integer by definition, so by Theorem 6.1(d) we may divide by n and obtain

$$-\frac{1}{kn} < \lambda - \frac{m}{n} < \frac{1}{kn}.$$

Futhermore, we know that since n is equal to $j - r$, it is less than k, and so the proof of Theorem 6.4 is complete.

Note that the number m/n is not necessarily in lowest terms. If $j - r$ and $a_j - a_r$ have no common factor, m/n is in lowest terms; otherwise it is not.

Problem Set 24

1. Following the statement of Theorem 6.4, there is an example given of the instance $\lambda = \sqrt{3}$ and $k = 8$. What values of m and n would have resulted if we had singled out the two numbers $0.464 \cdots$ and $0.392 \cdots$ in I_4, instead of the two numbers in I_6?

2. Apply the method given in the proof of Theorem 6.4 to each of the following cases, and so get values of m and n satisfying the inequalities of Theorem 6.4:

    (a) $\lambda = \sqrt{3}$, $k = 2$;   (h) $\lambda = \sqrt{2}$, $k = 8$;

    (b) $\lambda = \sqrt{3}$, $k = 4$;   (i) $\lambda = \sqrt{2}$, $k = 10$;

    (c) $\lambda = \sqrt{3}$, $k = 6$;   (j) $\lambda = \sqrt{2}$, $k = 14$;

    (d) $\lambda = \sqrt{3}$, $k = 10$;   (k) $\lambda = \pi$, $k = 2$;

    (e) $\lambda = \sqrt{2}$, $k = 2$;   (l) $\lambda = \pi$, $k = 4$;

    (f) $\lambda = \sqrt{2}$, $k = 4$;   (m) $\lambda = \pi$, $k = 6$;

    (g) $\lambda = \sqrt{2}$, $k = 6$;   (n) $\lambda = \pi$, $k = 8$.

6.5 Approximations to Within $1/n^2$

At the beginning of Section 6.4 we indicated the direction of our study, namely to try for better approximations of any irrational number λ. From the approximation of λ by m/n to within $1/2n$ for any n in

Theorem 6.3, we moved to approximation to within $1/kn$ for some $n \leq k$ in Theorem 6.4. Now we obtain approximations to within $1/n^2$.

THEOREM 6.5. *Given any irrational number λ, there are infinitely many rational numbers m/n in lowest terms such that*

$$-\frac{1}{n^2} < \lambda - \frac{m}{n} < \frac{1}{n^2}.$$

PROOF. First we observe that any rational number m/n satisfying the inequality of Theorem 6.4 automatically satisfies that of Theorem 6.5. The reason for this is that since n does not exceed k, from $k \geq n$ we may deduce, using Theorem 6.1, parts (d), (e), and (g), that

$$\frac{1}{k} \leq \frac{1}{n} \quad \text{and} \quad \frac{1}{kn} \leq \frac{1}{n^2}.$$

Hence any number which lies between $-1/kn$ and $1/kn$ must certainly lie in the range between $-1/n^2$ and $1/n^2$.

Next, we show that if any rational number m/n, not in lowest terms, satisfies the inequalities of the theorem, then the same rational number, in lowest terms, must also satisfy the appropriate inequalities. Let us write M/N as the form of m/n in lowest terms. We may presume that both n and N are positive, any negative sign being absorbed into the numerator. Hence we have

$$\frac{m}{n} = \frac{M}{N}, \quad 0 < N < n,$$

because the reduction to lowest terms does not alter the value of the fraction but does reduce the size of the denominator. It follows from Theorem 6.1 that

$$\frac{1}{n} < \frac{1}{N} \quad \text{and} \quad \frac{1}{n^2} < \frac{1}{N^2},$$

and so if λ satisfies

$$-\frac{1}{n^2} < \lambda - \frac{m}{n} < \frac{1}{n^2},$$

then it automatically satisfies

$$-\frac{1}{N^2} < \lambda - \frac{M}{N} < \frac{1}{N^2}.$$

In order to complete Theorem 6.5, all we must prove is that there are *infinitely many* rational numbers m/n, in lowest terms, that satisfy the inequalities. Suppose, on the contrary, that there were only a finite number of these fractions, say

$$\frac{m_1}{n_1}, \quad \frac{m_2}{n_2}, \quad \frac{m_3}{n_3}, \quad \ldots, \quad \frac{m_i}{n_i}.$$

Then consider the i numbers

$$\lambda - \frac{m_1}{n_1}, \quad \lambda - \frac{m_2}{n_2}, \quad \lambda - \frac{m_3}{n_3}, \quad \ldots, \quad \lambda - \frac{m_i}{n_i}.$$

These are all irrational by Theorem 4.1 of Chapter 4, and so none of them is zero. Some may be positive and others negative, and we choose an integer k so large that $1/k$ lies between 0 and all the positive numbers, and also so that $-1/k$ lies between 0 and all the negative numbers. We can do this because the larger we choose k, the closer $1/k$ and $-1/k$ get to 0. Thus we have chosen k so large that all the following inequalities are false:

$$-\frac{1}{k} < \lambda - \frac{m_1}{n_1} < \frac{1}{k},$$

(3)
$$-\frac{1}{k} < \lambda - \frac{m_2}{n_2} < \frac{1}{k},$$

$$\vdots$$

$$-\frac{1}{k} < \lambda - \frac{m_i}{n_i} < \frac{1}{k}.$$

With this value of k, we apply Theorem 6.4 and obtain a rational number m/n such that

$$-\frac{1}{kn} < \lambda - \frac{m}{n} < \frac{1}{kn}.$$

Now, this says that $\lambda - m/n$ lies between $-1/kn$ and $1/kn$, and so $\lambda - m/n$ must lie between $-1/k$ and $1/k$; in symbols,

$$-\frac{1}{k} < \lambda - \frac{m}{n} < \frac{1}{k}.$$

But since all the inequalities (3) are false, we conclude that m/n is different from each of the i numbers $m_1/n_1, m_2/n_2, \ldots, m_i/n_i$. Therefore we have obtained one more rational fraction to satisfy the inequalities of Theorem 6.5.

EXAMPLE. Find four rational approximations (in lowest terms) to the irrational number π, close enough to satisfy the inequalities of Theorem 6.5.

SOLUTION. First we observe that since $\pi = 3.14159 \cdots$,

$$-\frac{1}{12} < \pi - \frac{3}{1} < \frac{1}{12} \quad \text{and} \quad -\frac{1}{12} < \pi - \frac{4}{1} < \frac{1}{12}.$$

To find two others, we can use the method of Theorem 6.3 to get the closest rational numbers with denominators 2, 3, and so on:

$$\frac{6}{2}, \frac{9}{3}, \frac{13}{4}, \frac{16}{5}, \frac{19}{6}, \frac{22}{7}, \ldots .$$

We reject 6/2 and 9/3 because they are not in lowest terms, and we test the others in the inequalities of Theorem 6.5; for example,

$$-\frac{1}{36} < \pi - \frac{19}{6} < \frac{1}{36}. \qquad \text{(True!)}$$

Thus we are led to reject 13/4 and 16/5, and accept 19/6 and 22/7. So one set of answers to the question would be 3/1, 4/1, 19/6, and 22/7.

The rational number 22/7 is a very good approximation of π. There is no rational number with denominator between 1 and 56 that is closer to π. The rational number 179/57 is slightly closer to π than 22/7, but it does not satisfy the inequalities of Theorem 6.5. The rational number 355/113 satisfies the inequalities of Theorem 6.5 and is markedly closer to π than is 22/7. In fact it is accurate to six decimal places.

It is possible to prove the following stronger version of Theorem 6.5: *Given any irrational number λ, there are infinitely many rational numbers m/n in lowest terms such that*

$$-\frac{1}{n(n+1)} < \lambda - \frac{m}{n} < \frac{1}{n(n+1)}.$$

With the help of this theorem, the number 4/1 (which is a relatively poor approximation to π) can be eliminated in the above example.

In order to prove the stronger version of Theorem 6.5, we need a stronger version of Theorem 6.4. We shall sketch only the main steps and leave the details to the reader.

In the proof of Theorem 6.4, Dirichlet's pigeon-hole principle was used to argue that, given k numbers distributed over k intervals, either there is a number in the first interval or there is an interval containing at least two of these numbers. To get the stronger version of Theorem 6.4, we divide our unit interval into $k + 1$ subintervals and argue: Given k numbers distributed over $k + 1$ intervals, either there is a number in the first interval, or there is a number in the last interval, or there exists an interval containing at least two numbers. This use of the pigeon-hole principle enables us to substitute the stronger inequality,

$$-\frac{1}{n(k + 1)} < \lambda - \frac{m}{n} < \frac{1}{n(k + 1)},$$

for that now appearing in Theorem 6.4, without otherwise changing the statement. The proof of the stronger version of Theorem 6.5 is now immediate.

Problem Set 25

1. For a given irrational λ, prove that two of the "infinitely many rational numbers m/n" of Theorem 6.5 have $n = 1$, i.e., are integers.

2. Let λ be a given irrational number. Prove that, apart from one exception, any rational number that satisfies the inequalities of Theorem 6.5 automatically satisfies the inequalities of Theorem 6.3.

3. Find two rational numbers, which are not integers, that satisfy the inequalities of Theorem 6.5 for

   (a) $\lambda = \sqrt{2}$;   (b) $\lambda = \sqrt{3}$;   (c) $\lambda = \sqrt{5}$.

4. (a) Of the first five numbers in the sequence (1), which ones satisfy the inequalities of Theorem 6.3, with $\lambda = \sqrt{2}$?
   (b) Which ones satisfy the inequalities of Theorem 6.5?

5. (a) Of the first five numbers in the sequence (2), which ones satisfy the inequalities of Theorem 6.3. with $\lambda = \pi$?
   (b) Which ones satisfy the inequalities of Theorem 6.5?

*6. Prove that the statement of Theorem 6.5 is false in case $\lambda = 3/5$.

*7. (a) Let a/b and m/n be rational numbers, in lowest terms, with positive

denominators. Prove that they are unequal if $n > b$. Hence prove
that for $n > b$ the inequalities

$$-\frac{1}{bn} < \frac{a}{b} - \frac{m}{n} < \frac{1}{bn}$$

are false.

(b) Prove that the statement of Theorem 6.5 is false if λ is any fixed rational
number, say $\lambda = a/b$.

*8. Complete the proof of the stronger version of Theorem 6.5 (following the
sketch given just prior to this problem set); show that $\pi - 4/1$ and
$\pi - 19/6$ do not satisfy the stronger inequality, but that $\pi - 22/7$ does.

6.6 Limitations On Approximations

We proved in Theorem 6.3 that, corresponding to any irrational
number λ, there are infinitely many rational numbers m/n such that

$$-\frac{1}{2n} < \lambda - \frac{m}{n} < \frac{1}{2n}.$$

Then, in Theorem 6.5 we established that there are infinitely many m/n
such that

$$-\frac{1}{n^2} < \lambda - \frac{m}{n} < \frac{1}{n^2}.$$

Is it possible to prove that there are infinitely many m/n such that

$$-\frac{1}{2n^2} < \lambda - \frac{m}{n} < \frac{1}{2n^2}?$$

The answer is yes, although we shall not prove it here. In fact, there is
a famous theorem which states that there are infinitely many m/n
corresponding to any irrational number λ such that

$$-\frac{1}{\sqrt{5}\,n^2} < \lambda - \frac{m}{n} < \frac{1}{\sqrt{5}\,n^2},$$

and furthermore that $\sqrt{5}$ is the constant which yields the best possible
approximation of this kind. This means that if $\sqrt{5}$ is replaced by any
larger constant, the statement becomes false.

To give some idea as to how it is possible to prove that there is a
limit on the size of the constant, we establish the following result:

There are not infinitely many rational numbers m/n such that

(4) $$-\frac{1}{5n^2} < \sqrt{2} - \frac{m}{n} < \frac{1}{5n^2}.$$

In fact we prove that (4) is impossible for any integer n greater than 10.

The proof is indirect. We assume that (4) holds for some integers m and n, with $n > 10$. The inequality

$$-\frac{1}{5n^2} < \sqrt{2} - \frac{m}{n}$$

implies, for $n > 10$, that

(5) $$\frac{m}{n} < \sqrt{2} + \frac{1}{5n^2} < \sqrt{2} + \frac{1}{500} < 2.$$

On the other hand, the inequality

$$\sqrt{2} - \frac{m}{n} < \frac{1}{5n^2}$$

implies, for $n > 10$, that

(6) $$\frac{m}{n} > \sqrt{2} - \frac{1}{5n^2} > \sqrt{2} - \frac{1}{500} > 1.$$

Now, if we add m/n to the members of the inequalities (4), we get

(7) $$\frac{m}{n} - \frac{1}{5n^2} < \sqrt{2} < \frac{m}{n} + \frac{1}{5n^2}.$$

According to Theorem 6.1(e), each of these three parts can be squared and the inequalities retained, provided we prove that each part is positive. By (6) we see that

$$\frac{m}{n} - \frac{1}{5n^2} > 1 - \frac{1}{5n^2} > 1 - \frac{1}{500} > 0.$$

Hence all parts of (7) are positive, and so we square throughout to get

$$\left(\frac{m}{n} - \frac{1}{5n^2}\right)^2 < 2 < \left(\frac{m}{n} + \frac{1}{5n^2}\right)^2,$$

$$\frac{m^2}{n^2} - \frac{2m}{5n^3} + \frac{1}{25n^4} < 2 < \frac{m^2}{n^2} + \frac{2m}{5n^3} + \frac{1}{25n^4}.$$

Multiplying by n^2, we get

(8)
$$m^2 - \frac{2m}{5n} + \frac{1}{25n^2} < 2n^2 < m^2 + \frac{2m}{5n} + \frac{1}{25n^2}.$$

Now, by (5), we see that

(9)
$$m^2 + \frac{2}{5}\left(\frac{m}{n}\right) + \frac{1}{25n^2} < m^2 + \frac{2}{5}(2) + \frac{1}{25n^2}$$

$$< m^2 + \frac{4}{5} + \frac{1}{2500} < m^2 + 1.$$

On the other hand, by (5), we can write

(10) $\quad m^2 - \frac{2}{5}\left(\frac{m}{n}\right) + \frac{1}{25n^2} > m^2 - \frac{2}{5}\left(\frac{m}{n}\right) > m^2 - \frac{4}{5} > m^2 - 1.$

Applying (9) and (10) to (8), we obtain

$$m^2 - 1 < m^2 - \frac{2}{5}\left(\frac{m}{n}\right) + \frac{1}{25n^2} < 2n^2 < m^2 + \frac{2m}{5n} + \frac{1}{25n^2} < m^2 + 1,$$

$$m^2 - 1 < 2n^2 < m^2 + 1.$$

But $2n^2$ is an integer, so if it lies between the integers $m^2 - 1$ and $m^2 + 1$, it must equal m^2. Hence we conclude that

$$2n^2 = m^2, \qquad 2 = \frac{m^2}{n^2}, \qquad \sqrt{2} = \frac{m}{n},$$

and this is a contradiction, since $\sqrt{2}$ is irrational, while m and n were assumed to be integers.

Problem Set 26

1. (a) Prove that there are no rational numbers m/n, with $n > 10$, such that
$$-\frac{1}{5n^2} < \sqrt{3} - \frac{m}{n} < \frac{1}{5n^2}.$$

   (b) Find all rational numbers m/n satisfying these inequalities.

2. (a) Prove that there are no rational numbers m/n, with $n > 10$, such that
$$-\frac{1}{n^3} < \sqrt{2} - \frac{m}{n} < \frac{1}{n^3}.$$

   (b) Find all rational numbers m/n satisfying these inequalities.

3. (a) Prove that there are no rational numbers m/n, with $n > 10$, such that
$$-\frac{1}{n^3} < \sqrt{3} - \frac{m}{n} < \frac{1}{n^3}.$$

   (b) Find all rational numbers satisfying these inequalities.

6.7 Summary

We have established several results about how closely any irrational number λ can be approximated by infinitely many rational numbers m/n. The strongest theorem asserted that λ can be approximated to within $1/n^2$. Then in Section 6.6 we established a negative conclusion, namely that there do not exist infinitely many rationals m/n within $1/(5n^2)$ of $\sqrt{2}$. A similar negative conclusion applies to any algebraic number. It is true, but not proved here, that for any algebraic number λ, there do not exist infinitely many rational numbers m/n within $1/n^3$ of λ. This cannot be said of transcendental numbers in general; it is true of some, but not all, transcendental numbers. In the next chapter, we shall exhibit a number which can be approximated by infinitely many m/n not only to within $1/n^3$ but to within $1/n^4$, $1/n^{100}$, and, indeed, to within $1/n^j$ for any j the reader cares to name, however large. It will be proved that the number so exhibited is not algebraic, and thus we will have shown that there are such things as transcendental numbers. Up to now we have spoken of them without knowing that they even exist!

CHAPTER SEVEN

The Existence of Transcendental Numbers

How do we know that transcendental numbers exist? We shall answer this question in this final chapter. It is easy enough to exhibit a transcendental number; to prove that it is transcendental is quite another matter. The specific number whose transcendence we shall establish has the important feature that its decimal expansion consists mostly of zeros. We shall denote it by α, and its value is

$$\alpha = 0.1100010000\cdots,$$

where the ones occur in the decimal places numbered

$$1, \quad 2, \quad 6, \quad 24, \quad 120, \quad 720, \quad 5040, \quad \ldots,$$

that is to say, the decimal places numbered

$$1!, \quad 2!, \quad 3!, \quad 4!, \quad 5!, \quad 6!, \quad 7!, \quad \ldots.$$

The symbol $k!$, where k is a natural number, is read k *factorial* and denotes the product of all the natural numbers from 1 to k; thus

$$k! = 1 \cdot 2 \cdot 3 \cdot \cdots \cdot (k-2) \cdot (k-1) \cdot k.$$

All the digits in the decimal expansion of α are zero except those described above in the factorial number positions. Consequently α can be written as a sum of negative powers of 10; that is,

$$(1) \qquad \alpha = 10^{-1!} + 10^{-2!} + 10^{-3!} + 10^{-4!} + 10^{-5!} + \cdots$$

$$= 10^{-1} + 10^{-2} + 10^{-6} + 10^{-24} + 10^{-120} + \cdots$$

$$= 0.1 + 0.01 + 0.000001 + \cdots.$$

This number α is called a Liouville number, after the French mathematician who first demonstrated that transcendental numbers exist.

What concrete property of the transcendental number α can we use in order to prove that α is not algebraic? The answer is that α can be approximated by infinitely many rational numbers m/n, not only to within $1/n^2$ (this could be done for any irrational number, see Chapter 6) but to within $1/n^3$, $1/n^4$, and, in fact, to within $1/n^r$, where r is any positive number whatever. No algebraic number has this property. If λ is any irrational number, it can be approximated to within $1/n^2$ by infinitely many rational m/n, as we saw in Theorem 6.5. But if λ is algebraic, it cannot be approximated by infinitely many m/n more closely, not to within $1/n^3$, or even to within $1/n^{2.1}$; within $1/n^2$ is the best possible among all $1/n^r$. To find this kind of a result about algebraic numbers was for many years an outstanding unsolved problem. It was settled in 1955 by the British mathematician K. F. Roth who, for this ingenious work, was awarded a Fields medal in 1958 at the International Congress of Mathematicians in Edinburgh, Scotland. The result is known as the Thue–Siegel–Roth theorem because A. Thue and C. L. Siegel proved certain underlying results upon which Roth's work was based.

As we said, proving the transcendence of α is a more difficult matter than merely writing the decimal expansion of α. We shall use the ideas on inequalities from Sect. 6.1. We also need the concept of absolute value. Perhaps the reader is familar with this concept, but in case he is not, we give a brief introduction to this topic and also to the factor theorem.

7.1 Some Algebraic Preliminaries

Any real number a is either positive, negative, or zero. For every such number a, we shall define the "absolute value of a," denoted by the symbol $|a|$.† If a is positive or zero, we define the absolute value of a by the equation $|a| = a$. If a is negative, the definition is $|a| = -a$. For example,

$$|0| = 0, \quad |7| = 7, \quad |-4| = 4, \quad |-6| = 6,$$
$$|3| = 3, \quad |-5| = 5, \quad |-1000| = 1000.$$

Instead of separating the definition into cases in which a is positive,

---

† For a detailed treatment of the concept of absolute value, see Chapter III of the monograph by Edwin Beckenbach and Richard Bellman, in this series.

zero, or negative, we could define the absolute value of a by the single equation

(2) $$|a| = \sqrt{a^2},$$

because of the convention that $\sqrt{a^2}$ never designates a negative value.

One basic result is that if two numbers are equal, so are their absolute values. In symbols, if $a = b$, then $|a| = |b|$. Another simple consequence of our definition (2) is that a and $-a$ have the same absolute value, whatever may be the value of a. In symbols, $|a| = |-a|$.

Another important result is that $|ab| = |a| \cdot |b|$. We can prove this easily by using (2), as follows:

$$|a| = \sqrt{a^2}, \qquad |b| = \sqrt{b^2}, \qquad |ab| = \sqrt{a^2 b^2} = \sqrt{a^2} \cdot \sqrt{b^2},$$

so that

$$|ab| = |a| \cdot |b|.$$

Next, how is $|a + b|$ related to the sum $|a| + |b|$? We shall show that $|a + b| \leq |a| + |b|$. To prove this result, known as the *triangle inequality* in the wider setting of complex numbers, we separate the problem into cases. If a and b are both positive, then

$$|a + b| = a + b, \qquad |a| = a, \qquad |b| = b,$$

so that

$$|a + b| = |a| + |b|.$$

If a and b are both negative, then

$$|a + b| = -a - b, \qquad |a| = -a, \qquad |b| = -b,$$

so that, again,

$$|a + b| = |a| + |b|.$$

If a and b are of mixed signs, one positive and the other negative, then $a + b$ involves, in effect, some cancellation, and we can say that

$$|a + b| \text{ is less than the larger of } |a| \text{ and } |b|.$$

It follows that $|a + b| < |a| + |b|$.

If one of the numbers is zero, for example if $b = 0$, then

$$|a + b| = |a + 0| = |a|, \qquad |b| = |0| = 0,$$

so

$$|a + b| = |a| + |b|.$$

In summary, we see that in all cases, we have either

$$|a + b| = |a| + |b| \qquad \text{or} \qquad |a + b| < |a| + |b|.$$

All these results on absolute values are collected in the following theorem for convenience.

THEOREM 7.1. *For all real numbers a and b, we have:*

(1) *If $a = b$, then $|a| = |b|$;*
(2) $|a| = |-a|$;
(3) $|ab| = |a| \cdot |b|$;
(4) $|a + b| \leq |a| + |b|$.

Next, we prove the *factor theorem* from algebra. Actually we prove it in a special setting for our own later purposes.

THEOREM 7.2. *Let $f(x)$ be a polynomial with integer coefficients, and let the rational number β be a root of $f(x) = 0$. Then $x - \beta$ is a factor of $f(x)$; that is, there is a polynomial $q(x)$ such that $f(x) = (x - \beta) q(x)$. Furthermore, $q(x)$ has rational coefficients, and has degree one less than the degree of $f(x)$.*

PROOF. If we divide $x - \beta$ into $f(x)$, there results a quotient $q(x)$ and a remainder, say r. Since the degree of the remainder is always less than that of the divisor (which, in our case, is the first degree polynomial $x - \beta$), we see that r is a constant independent of x. It follows that

$$f(x) = (x - \beta) q(x) + r,$$

and since the steps in the division process are so-called rational operations, we see that $q(x)$ must have rational coefficients. The equation above is an identity in x, so we can now replace x by β to get $f(\beta) = r$. However, $f(\beta) = 0$ because β is a root of $f(x) = 0$. Hence $r = 0$. Thus the division $f(x)$ by $x - \beta$ gives a zero remainder, and so $f(x) = (x - \beta) q(x)$. Finally, whatever the degree of $f(x)$, we see that $q(x)$ has degree one less.

Problem Set 27

1. Write the values of $|2|$, $|-2|$, $|-8|$, and $|10^{-1}|$.

2. In the text it was established if $a = b$, then $|a| = |b|$. Is the converse true?

3. Prove that $|a + b + c| \leq |a| + |b| + |c|$.

4. (a) Prove that $|x + 7| = x + 7$ if $x \geq -7$, but $|x + 7| = -x - 7$ if $x \leq -7$.
   (b) Give a similar analysis of $|x - 7| = x - 7$.

5. For what values of x, if any, do the following equations hold?
   (a) $|x + 7| = 5 + |x|$;   (c) $|x + 7| + |x - 7| = |x| + 7$;
   (b) $|x| = |x - 4|$;   (d) $|2x| = 2 |x|$.

6. Prove that the inequalities of Theorem 6.5 of Chapter 6,

$$-\frac{1}{n^2} < \lambda - \frac{m}{n} < \frac{1}{n^2},$$

can be written in the form

$$\left| \lambda - \frac{m}{n} \right| < \frac{1}{n^2}.$$

7. Prove that $8! = 8(7!)$; also prove that $(j + 1)! = (j + 1)(j!)$.

8. Prove that $(j + 1)! - j! = j(j!)$.

9. Verify that $3/2$ is a root of $2x^4 - 13x^3 + 27x^2 - 4x - 21 = 0$. Then denote this polynomial by $f(x)$ and verify Theorem 7.2 by computing the quotient $q(x)$ in the division of $f(x)$ by $x - 3/2$.

7.2 An Approximation to α

The underlying reason for the transcendence of α is that it can be approximated exceptionally well by certain rational numbers. This we now demonstrate. A good rational approximation to α can be obtained by taking a finite number of terms from the series (1) that defines α. We define β as the sum of the first j terms of α as given in (1); that is,

$$(3) \qquad \beta = 10^{-1!} + 10^{-2!} + 10^{-3!} + \cdots + 10^{-j!}.$$

The value of the integer j will be specified later. We observe that β is rational because it can be written as a sum of fractions, whose denominators are powers of 10;

$$\beta = \frac{1}{10^{1!}} + \frac{1}{10^{2!}} + \frac{1}{10^{3!}} + \cdots + \frac{1}{10^{j!}}.$$

These fractions can all be written with common denominator $10^{j!}$, and so they can be added to give a single fraction,

$$(4) \qquad\qquad \beta = \frac{t}{10^{j!}},$$

where the numerator t denotes some *integer* whose exact value is immaterial.

The rational number β is very close to α. From eqs. (1) and (3), we see that

$$\alpha - \beta = 10^{-(j+1)!} + 10^{-(j+2)!} + 10^{-(j+3)!} + \cdots.$$

The decimal expansion of $\alpha - \beta$, like that of α itself, consists entirely of zeros and ones. The digit 1 appears first in the $(j+1)!$ place, then in the $(j+2)!$ place, and so on. Thus the number $\alpha - \beta$ is less than

$$0.000000 \cdots 0000002,$$

where all digits are zeros except the digit 2 in the $(j+1)!$ place. Another way of saying this is

$$(5) \qquad\qquad \alpha - \beta < \frac{2}{10^{(j+1)!}}.$$

We shall need some other simple inequalities involving α and β. Since α and β are positive, so are all powers of α and β. Furthermore, since $\alpha < 1$ and $\beta < 1$, we see that $\alpha^r < 1$, that $\beta^s < 1$, and that $\alpha^r \beta^s < 1$ for any positive integers r and s, and so we have

$$(6) \qquad 0 < \alpha^r < 1, \qquad 0 < \beta^s < 1, \qquad 0 < \alpha^r \beta^s < 1.$$

7.3 The Plan of the Proof

'To prove that α is transcendental, we shall assume exactly the reverse, namely that α is algebraic, and we shall obtain a contradiction. The assumption that α is algebraic means that α satisfies some algebraic equation with integer coefficients. Among all the algebraic equations with integer coefficients satisfied by α, select one of lowest degree, say

$$(7) \quad c_n x^n + c_{n-1} x^{n-1} + c_{n-2} x^{n-2} + \cdots + c_2 x^2 + c_1 x + c_0 = 0.$$

For brevity we shall write $f(x)$ for the polynomial on the left side of

(7). This polynomial $f(x)$ will play a central role throughout the rest of the chapter. The basic assumptions about $f(x)$ to be kept in mind are these:

(1) it has integer coefficients;
(2) the number α is a root of $f(x) = 0$, so that $f(\alpha)$ is identically zero [where by $f(\alpha)$ we mean the result of replacing x by α in $f(x)$];
(3) the number α is a root of no equation of degree less than n, with integer coefficients.

The number $f(\beta)$, obtained by replacing x by β in $f(x)$, will also play a central role in the analysis.

The idea of the proof is this. We will look at the number $f(\alpha) - f(\beta)$ [or $-f(\beta)$ which amounts to the same thing, since $f(\alpha) = 0$] in two different ways. One way of looking at $-f(\beta)$ is as a polynomial in β, with integer coefficients. Since β is rational, $-f(\beta)$ is also rational, and we shall see that its absolute value is relatively large. Another way of looking at $f(\alpha) - f(\beta)$ is as the difference of two polynomials, and we shall show in the next section that this difference has the same order of magnitude as $\alpha - \beta$, which is relatively small [see eq. (5)]. Thus, by assuming that α is algebraic, we shall deduce two conflicting orders of magnitude for $f(\alpha) - f(\beta)$ and so establish a contradiction.

We shall prepare the way for this in the next section by showing that $f(\beta)$ is not zero, and that $f(\alpha) - f(\beta)$ has the same order of magnitude as $\alpha - \beta$.

Problem Set 28

1. Verify the identities:
   (a) $\alpha^4 - \beta^4 = (\alpha - \beta)(\alpha^3 + \alpha^2\beta + \alpha\beta^2 + \beta^3)$;
   (b) $\alpha^5 - \beta^5 = (\alpha - \beta)(\alpha^4 + \alpha^3\beta + \alpha^2\beta^2 + \alpha\beta^3 + \beta^4)$;
   (c) $\alpha^6 - \beta^6 = (\alpha - \beta)(\alpha^5 + \alpha^4\beta + \alpha^3\beta^2 + \alpha^2\beta^3 + \alpha\beta^4 + \beta^5)$.

2. Write an identity expressing $\alpha^7 - \beta^7$ as $\alpha - \beta$ multiplied by a polynomial of degree 6.

3. Prove that an algebraic number is a root of infinitely many algebraic equations with integer coefficients.

7.4 Properties of Polynomials

THEOREM 7.3. *The number β is not a root of eq. (7); that is, $f(\beta) \neq 0$.*

PROOF. If β were a root of (7) then by Theorem 7.2, $x - \beta$ would be a factor of $f(x)$, say

$$f(x) = (x - \beta)\, q(x).$$

Also by Theorem 7.2, $q(x)$ has rational coefficients, and its degree is one less than the degree of $f(x)$. Now, since α is a root of $f(x) = 0$, we have

$$f(\alpha) = (\alpha - \beta)\, q(\alpha) = 0.$$

But this product is zero only if at least one of the factors is zero. The factor $\alpha - \beta$ is not zero because α is different from β. Hence $q(\alpha) = 0$; that is, α is a root of $q(x) = 0$, and $q(x)$ is of degree $n - 1$. If we denote by k the product of all the denominators of the rational coefficients of $q(x)$, then the product $kq(x)$ has integer coefficients and α is a root of $kq(x) = 0$. But this contradicts the fact that α satisfies no equation of degree less than n, with integer coefficients. Since the assumption that $f(\beta) = 0$ has led to a contradiction, we conclude that $f(\beta) \neq 0$.

Next, following the outline given in the last section, we show that $|f(\alpha) - f(\beta)|$ is of the same order of magnitude as $|\alpha - \beta|$, which is very small (see Section 7.2).

THEOREM 7.4. *There is a number N, dependent only on the coefficients of $f(x)$ and its degree, such that*

$$|f(\alpha) - f(\beta)| < N\,(\alpha - \beta).$$

PROOF. The number N is defined by the equation

$$(8) \quad N = n|c_n| + (n - 1)\,|c_{n-1}| + (n - 2)\,|c_{n-2}| + \cdots + 2|c_2| + |c_1|.$$

Observe, in particular, that N is independent of the integer j used in the definition of β.

In the course of the proof, we shall also need the factoring of $\alpha^k - \beta^k$ and an inequality satisfied by $\alpha^k - \beta^k$. The factoring is given by

$$(9) \quad \alpha^k - \beta^k = (\alpha - \beta)\,(\alpha^{k-1} + \alpha^{k-2}\beta + \alpha^{k-3}\beta^2 + \cdots$$
$$+ \alpha^2\beta^{k-3} + \alpha\beta^{k-2} + \beta^{k-1}),$$

where k is any positive integer. This factoring can be verified by multiplying out the right-hand side of (9):

$$\alpha(\alpha^{k-1} + \alpha^{k-2}\beta + \cdots + \alpha\beta^{k-2} + \beta^{k-1})$$
$$= \alpha^k + \alpha^{k-1}\beta + \cdots + \alpha^2\beta^{k-2} + \alpha\beta^{k-1}$$

and

$$\beta(\alpha^{k-1} + \alpha^{k-2}\beta + \cdots + \alpha\beta^{k-2} + \beta^{k-1})$$
$$= \alpha^{k-1}\beta + \alpha^{k-2}\beta^2 + \cdots + \alpha\beta^{k-1} + \beta^k.$$

When we subtract these two equations, we observe that all terms, except the first term of the first equation and the last term of the second equation, cancel. Thus only $\alpha^k - \beta^k$ remains.

Looking at the right side of eq. (9), we see that each of the terms α^{k-1}, $\alpha^{k-2}\beta$, etc. is less than 1, by the inequalities (6). But since there are exactly k of these terms and since $\alpha - \beta$ is positive, we can write

$$(10) \quad \alpha^k - \beta^k < (\alpha - \beta)(1 + 1 + 1 + \cdots + 1 + 1 + 1) = k(\alpha - \beta).$$

Now we compute $f(\alpha)$ and $f(\beta)$ using eq. (7) and subtract $f(\beta)$ from $f(\alpha)$. Thus we obtain

$$f(\alpha) - f(\beta) = c_n(\alpha^n - \beta^n) + c_{n-1}(\alpha^{n-1} - \beta^{n-1}) + \cdots + c_1(\alpha - \beta).$$

Next we use the identity (9) to take out the common factor $\alpha - \beta$ from all the terms on the right. This leads to

$$f(\alpha) - f(\beta) = (\alpha - \beta)\,[c_n(\alpha^{n-1} + \alpha^{n-2}\beta + \cdots + \alpha\beta^{n-2} + \beta^{n-1})$$
$$+ c_{n-1}(\alpha^{n-2} + \alpha^{n-3}\beta + \cdots + \alpha\beta^{n-3} + \beta^{n-2})$$
$$+ \cdots + c_1].$$

Taking absolute values and using Theorem 7.1 and inequality (10), we get

$$|f(\alpha) - f(\beta)| < |\alpha - \beta|\,[n|c_n| + (n-1)\,|c_{n-1}| + \cdots + |c_1|].$$

Observing that $|\alpha - \beta| = \alpha - \beta$ and using the defining equation (8) for the number N, we finally have $|f(\alpha) - f(\beta)| < N(\alpha - \beta)$ and the theorem is proved.

7.5 The Transcendence of α

We now complete the proof that the number α defined by eq. (1) is transcendental. First, we look at $f(\alpha) - f(\beta)$ in another way.

THEOREM 7.5. *The number*

(11) $$|f(\alpha) - f(\beta)| \cdot 10^{n \cdot j!}$$

is a positive integer, no matter what value is assigned to the positive integer j.

PROOF. Since $f(\alpha) = 0$, the number under discussion can be written as

$$|-f(\beta)| \cdot 10^{n \cdot j!} \quad \text{or} \quad |f(\beta)| \cdot 10^{n \cdot j!}.$$

From eqs. (7) and (4) we see that

$$f(\beta) = c_n \beta^n + c_{n-1} \beta^{n-1} + c_{n-2} \beta^{n-2} + \cdots + c_1 \beta + c_0$$

$$= \frac{c_n t^n}{10^{n \cdot j!}} + \frac{c_{n-1} t^{n-1}}{10^{(n-1)j!}} + \frac{c_{n-2} t^{n-2}}{10^{(n-2)j!}} + \cdots + \frac{c_1 t}{10^{j!}} + c_0.$$

Multiplying by $10^{n \cdot j!}$ we have

$$f(\beta) \cdot 10^{n \cdot j!}$$
$$= c_n t^n + c_{n-1} t^{n-1} 10^{j!} + c_{n-2} t^{n-2} 10^{2 \cdot j!} + \cdots + c_1 t 10^{(n-1)j!} + c_0 10^{n \cdot j!},$$

and the right-hand side is an integer. This integer cannot be zero, because $f(\beta) \neq 0$ by Theorem 7.3. Taking absolute values, we see that

$$|f(\beta) \cdot 10^{n \cdot j!}| \quad \text{or} \quad |f(\beta)| \cdot 10^{n \cdot j!}$$

is a positive integer, and thus the theorem is proved.

We shall now get an outright contradiction to Theorem 7.5 by showing that the number given by (11) lies between 0 and 1. In order to do this, we must choose the integer j, which was used in the definition of β, to satisfy

(12) $$\frac{2N \cdot 10^{n \cdot j!}}{10^{(j+1)!}} < 1.$$

Can this be done? It can, because this inequality is equivalent to

$$\frac{2N}{10^{(j+1)! - n \cdot j!}} < 1,$$

where the exponent in the denominator can be written as

$$(j+1)! - n \cdot j! = (j+1)j! - n \cdot j! = (j+1-n)j!.$$

This exponent can be made as large as we please for fixed n, by taking j very large. Now n and N are fixed by the equations (7) and (8); but since j depends neither on n nor on N, we can take j so large that (12) is satisfied.

Next we show that the number given by (11) lies between 0 and 1, by using Theorem 7.4 and inequality (5), thus:

$$|f(\alpha) - f(\beta)| \cdot 10^{n \cdot j!} < N(\alpha - \beta) \cdot 10^{n \cdot j!}$$

$$< \frac{2N \cdot 10^{n \cdot j!}}{10^{(j+1)!}}$$

$$< 1,$$

where in the last step we used (12). Of course the number in (11) is positive, because of Theorem 7.3.

Hence we have a contradiction, and we conclude that α cannot satisfy any equation of the form (7). So α is a transcendental number.

7.6 Summary

In this chapter we have answered the question: "Are there any transcendental numbers?" by actually exhibiting a Liouville number and by proving that it is transcendental, i.e., not algebraic.

Let us recapitulate the entire proof, since the details may have obscured the argument. We said at the beginning of the chapter that the central idea is that the number

$$\alpha = 10^{-1!} + 10^{-2!} + 10^{-3!} + 10^{-4!} + \cdots$$

can be approximated very closely by rational numbers. This fact is stated in the inequality (5) which says, in effect, that $\alpha - \beta$ is very small compared to β. We recall that β is a rational number with denominator $10^{j!}$ [see eq. (4)], but that $\alpha - \beta$ is of the order $10^{-(j+1)!}$. In Theorem 7.4, this small order of magnitude was extended from $\alpha - \beta$ to $f(\alpha) - f(\beta)$, where $f(x)$ is a polynomial with integer coefficients which, for $x = \alpha$, allegedly vanishes.

On the other hand, by considering $f(\alpha) - f(\beta)$ in quite a different way in Theorem 7.5, we showed that the magnitude of $f(\alpha) - f(\beta)$ is larger than the earlier estimate. (The factor $10^{n \cdot j!}$ in Theorem 7.5 plays no essential role; it is present in order to place the two orders of magnitude of $f(\alpha) - f(\beta)$ in sharp contrast.) This was done by recognizing that $f(\alpha) - f(\beta)$ is simply $-f(\beta)$, and $f(\beta)$ is a rational number with denominator $10^{n \cdot j!}$. Hence the assumption that α satisfies $f(x) = 0$ enables us to prove that $f(\alpha) - f(\beta)$ is much larger than the earlier calculation showed. This contradiction establishes the fact that α is transcendental.

Proof That There Are Infinitely Many Prime Numbers

The argument made here is a so-called indirect proof, otherwise known as proof by contradiction, or *reductio ad absurdum*. In this type of proof, we assume that the proposition is false and then derive a contradiction from this assumption. Thus, in the case of the present proposition, we assume that there is only a finite number of primes.

Next we devise a system of notation for the primes. There being only a finite number, let us denote them by

$$p_1, p_2, p_3, \ldots, p_k.$$

This notation means that there are exactly k primes, where k is some natural number. If we regard these primes as being listed in order of size, then of course $p_1 = 2$, $p_2 = 3$, $p_3 = 5$, $p_4 = 7$, and so on. Nevertheless, in this proof, it is more convenient to use the notation p_1, p_2, p_3, etc. rather than 2, 3, 5, etc.

Since every natural number can be factored into primes, we observe that every natural number must be divisible by at least one of the primes

$$p_1, p_2, p_3, \ldots, p_k,$$

because by our assumption there are no primes other than these. But consider the natural number n which is obtained by multiplying all the primes and then adding 1:

$$n = p_1 p_2 p_3 \cdots p_k + 1.$$

This number n is not divisible by p_1, because if we divide p_1 into n we get a quotient and a remainder with values,

$$\text{quotient} = p_2 p_3 \cdots p_k, \qquad \text{remainder} = 1.$$

115

If n were divisible by p_1, the remainder would be 0; thus n is not divisible by p_1.

A similar argument shows that n is not divisible by p_2, or p_3, or $p_4, \ldots$, or p_k.

We have exhibited a number n which is divisible by no prime whatever, and this is an absurd situation. So the assumption that there were only finitely many primes has led to a logical contradiction, and consequently this assumption must have been false. Hence there are infinitely many primes.

APPENDIX B

Proof of the Fundamental Theorem of Arithmetic

It is proved in this appendix that *every natural number other than 1 can be factored into primes in only one way, except for the order of the factors.* It is understood that any natural number which is itself a prime, such as 23, is a "factoring into primes" as it stands. Now the result can be readily checked for small natural numbers. For example, 10 can be factored as $2 \cdot 5$, and we know from experience that there is no other factoring. The same is true of all numbers up to 10:

$$2 = 2$$
$$3 = 3$$
$$4 = 2 \cdot 2$$
$$5 = 5$$
$$6 = 2 \cdot 3$$
$$7 = 7$$
$$8 = 2 \cdot 2 \cdot 2$$
$$9 = 3 \cdot 3$$
$$10 = 2 \cdot 5$$

This list could be continued, but such a listing, however long, could not be regarded as a proof. For, after all, there are infinitely many natural numbers, and we cannot check the factoring of them all.

So we must turn to a mathematical argument. The natural numbers from 2 to 10 have been listed, each with its unique factorization. Now, either this list can be extended indefinitely so that there is unique factorization for every natural number, or at some place in the continued listing the unique factorization property breaks down. These are the only two possibilities. It is the first of these two possibilities that we propose to establish, and we shall do it by an indirect argument.

117

We assume that the second possibility holds, i.e., that at some place in the listing of natural numbers the unique factorization property breaks down, and show that this leads to a contradiction.

Before carrying out the details of this rather long argument, we give a brief sketch to guide the reader.

We shall denote the first integer which can be factored into primes in more than one way by m, and we shall write two different prime factorizations for m. In Part I of the proof, we shall show that none of the primes in one factorization of m occurs in the other. Having established that, if m indeed had two different factorizations, all primes in one would be different from all primes in the other, we finally construct, in Part II of the proof, a number n which is smaller than m and also has two different factorizations into primes. This contradicts the assumption that m was the smallest integer having two different prime factorizations, and thus completes the proof.

Let us denote by m the first integer which can be factored into primes in more than one way. In other words we assume that every natural number smaller than m has the unique factorization property, whereas m has more than one factoring. Thus we know that there are at least two different factorizations of m, for which we write

$$m = p_1 p_2 p_3 \cdots p_r \quad \text{and} \quad m = q_1 q_2 q_3 \cdots q_s.$$

What do we mean by this notation? We mean that m can be factored into primes p_1, p_2, p_3, and so on, as far as p_r, and that there is also another way of factoring m into the primes q_1, q_2, q_3, and so on, to q_s. Why not q_1, q_2, q_3, and so on, to q_r? Because we cannot presume that the number of prime factors in the two factorings will be the same; they might be different for all we know.

The notation needs further explanation. We do not mean, as we did in Appendix A, that p_1 is just another label for the prime 2, p_2 another label for the prime 3, and so on. Not at all. We do not know whether or not the prime 2 is in the batch p_1, p_2, $\ldots$, p_r. So p_1 may be 2, or it may be 23, or it may be 47, or it may be none of these. It is simply some prime. Similarly, p_2 is just some prime. It may be the same prime as p_1, or it may not. All that we are presuming is that the natural number m can be factored into primes in two different ways.

PART I OF THE PROOF. Now, the first thing that we can show is that the primes p_1, p_2, $\ldots$, p_r in the first batch are entirely different from the primes q_1, q_2, $\ldots$, q_s in the second batch. In other words, if the prime 7 occurs in the first batch it cannot occur in the second batch. Since this is not at all obvious, we must give an argument. If the two batches had a prime in common, we could arrange the notation so that the common prime would be the first one in each batch, thus $p_1 = q_1$. (We may do this because in each factorization the primes can be thought of in any order.) Now, since $p_1 = q_1$, we may as well write p_1 in place of q_1, so that the two factorizations are

$$m = p_1 p_2 p_3 \cdots p_r \quad \text{and} \quad m = p_1 q_2 q_3 \cdots q_s.$$

Dividing these equations by p_1, we obtain

$$\frac{m}{p_1} = p_2 p_3 \cdots p_r \quad \text{and} \quad \frac{m}{p_1} = q_2 q_3 \cdots q_s.$$

Now we have two different factorings for the natural number m/p_1 because we began with two different factorings for m. But this is impossible because m was the smallest number having more than one factoring, and m/p_1 is smaller than m.

PART II OF THE PROOF. Thus we have established that the primes $p_1, p_2, \ldots, p_r$ in the first factorization of m are entirely different from the primes $q_1, q_2, \ldots, q_s$ in the second. In particular, we know that p_1 is not equal to q_1; in mathematical symbols, $p_1 \neq q_1$. We shall presume that p_1 is the smaller of the two; that is, $p_1 < q_1$. We have a right to presume this because the notation is entirely symmetric between the two batches of primes. Thus, if we can complete the proof in the case $p_1 < q_1$, a symmetric proof with the p's and q's interchanged must apply in an analogous way to the case $p_1 > q_1$.

Presuming, then, that $p_1 < q_1$, we shall exhibit a number smaller than m having two different factorings. This will complete the proof because we shall have contradicted the assumption we made at the outset, that m was the smallest number with more than one factoring. A natural number that will meet the stated specifications is

$$n = (q_1 - p_1) \, q_2 q_3 q_4 \cdots q_s.$$

Note how n is constructed: it is the product of $q_1 - p_1$ and the primes $q_2, q_3, \ldots, q_s$. It can be written as a difference,

$$n = q_1 q_2 q_3 \cdots q_s - p_1 q_2 q_3 \cdots q_s$$

or

$$n = m - p_1 q_2 q_3 \cdots q_s,$$

and since $p_1q_2q_3 \cdots q_s$ is a positive number, this shows that n is smaller than m.

Finally, we establish that the natural number n has two different factorings. To do this, we look at the form in which n was introduced, namely

$$n = (q_1 - p_1) \, q_2q_3 \cdots q_s.$$

Each of the factors $q_2, q_3, \ldots, q_s$ is a prime, but the first factor, $q_1 - p_1$, is not necessarily a prime. If $q_1 - p_1$ were to be factored into primes, we would have a factoring of n into primes which *would not include the prime p_1 as one of the factors*. To see this, we observe first that the primes $q_2, q_3, \ldots, q_s$ do not have p_1 among them, as shown in Part I of the proof. Second, regardless of how $q_1 - p_1$ is factored into primes, the prime p_1 could not be present; for, if p_1 were a factor in the prime factorization of $q_1 - p_1$, then p_1 would be a divisor of $q_1 - p_1$. That is, the equation

$$q_1 - p_1 = p_1b,$$

where b is the quotient in the division process, would hold. But this would lead to the equations

$$q_1 = p_1 + p_1b \quad \text{and} \quad q_1 = p_1 \, (1 + b),$$

and the latter can be interpreted as stating that p_1 is a divisor of q_1, which is impossible since no prime can be a divisor of another prime.

Next, we show that n can also be factored in another way so that p_1 *is one of the prime factors*. To do this, we return to an earlier equation

$$n = m - p_1q_2q_3 \cdots q_s,$$

replace m by its form

$$m = p_1p_2p_3 \cdots p_r$$

and so obtain

$$n = p_1p_2p_3 \cdots p_r - p_1q_2q_3 \cdots q_s$$
$$= p_1(p_2p_3 \cdots p_r - q_2q_3 \cdots q_s).$$

The part in parentheses is not necessarily a prime; but if we factored it into primes, we would have a prime factorization of n which includes the prime p_1. Thus we have exhibited two factorings of n or, rather, two procedures for obtaining factorings of n, one without the prime p_1 among the factors and the other with. In other words, the number n, which is smaller than m, has two different prime factorizations. This completes the proof.

APPENDIX C

Cantor's Proof of the Existence of Transcendental Numbers

In Chapter 7 we established the existence of transcendental numbers by exhibiting one. In this appendix, we shall give an independent proof of their existence by an entirely different method, at the same time showing that there are infinitely many transcendental numbers. In fact we establish that in a certain sense there are more transcendental than algebraic numbers.

At the outset, let us make clear that we are confining our attention to *real* algebraic numbers and *real* transcendental numbers. The roots of $x^2 + 1 = 0$, for example, are algebraic, but not *real* algebraic, numbers. The results we state and their proofs are valid in the complex case also, but we avoid a few minor complications by restricting our attention to real numbers.

By a set S we mean any collection of definite, well-distinguished objects. These objects are called the members of the set S, or the elements of S. A set S may be finite as, for example, the set of prime numbers less than 20,

$$S = \{2, 3, 5, 7, 11, 13, 17, 19\};$$

or S may be infinite as, for example, the set of all natural numbers

$$S = \{1, 2, 3, 4, 5, 6, \ldots\}.$$

An infinite set is said to be countable (or denumerable) if its members can be written as a sequence

$$a_1, a_2, a_3, a_4, \ldots,$$

so that every element of the set is to be found in the sequence. For example, the set of even natural numbers can be written as

$$2, 4, 6, 8, 10, 12, \ldots,$$

so that the nth term in the sequence is $2n$, and hence this is a countable set.

The set of all integers is countable, because it can be written as a sequence

$$0, \quad 1, -1, \quad 2, -2, \quad 3, -3, \quad 4, -4, \ldots.$$

It can be written as a sequence in other ways, but any one way is sufficient to show that the set is countable.

It is not necessary that we know any specific formula for the nth term of a sequence in order to conclude that a set is countable. For example, the set of primes

$$2, 3, 5, 7, 11, 13, 17, 19, \ldots$$

is countable, even though we do not know the precise value of the hundred-millionth prime. It is enough to know that there is such a prime, so that we can conceive of a sequential order for the entire set.

Next, we establish that the set of all rational numbers is countable. We note that any rational number is a root of a linear equation $ax + b = 0$, with integer coefficients a and b. Futhermore we restrict a to be positive, without any loss of generality. For example, the rational number $3/5$ is a root of $5x - 3 = 0$. We say that the equation $ax + b = 0$ has *index*

$$1 + a + |b|,$$

so that the index of an equation is a positive integer. For example, the equation $5x - 3 = 0$ has index 9. There is no equation with index 1, and only one with index 2, namely $x = 0$. Table C 1 includes all linear equations with indices up to 5. The rational numbers, in order of size, introduced by the equations of Table C 1 can also be written in tabular form, as shown in Table C 2.

It is clear that for any index j there is only a finite number of linear equations. In fact, there are $2j - 3$ equations with index j (the precise number really has no significance). So with each increasing index only

TABLE C 1

Index	Equations
2	$x = 0$
3	$2x = 0, \ x + 1 = 0, \ x - 1 = 0$
4	$3x = 0, \ 2x + 1 = 0, \ 2x - 1 = 0, \ x + 2 = 0, \ x - 2 = 0$
5	$4x = 0, \ 3x + 1 = 0, \ 3x - 1 = 0, \ 2x + 2 = 0, \ 2x - 2 = 0,$ $x + 3 = 0, \ x - 3 = 0$

TABLE C 2

Index	Rational Numbers Introduced
2	0
3	$-1, +1$
4	$-2, -\frac{1}{2}, \frac{1}{2}, 2$
5	$-3, -\frac{1}{3}, \frac{1}{3}, 3$

a finite number of new rational numbers are introduced. Hence we can write the rational numbers as a sequence

$$0, -1, \quad 1, -2, -\frac{1}{2}, \quad \frac{1}{2}, \quad 2, -3, -\frac{1}{3}, \quad \frac{1}{3}, \quad 3, \ldots,$$

by listing the roots of the equation of index 2, then the roots of all the equations of index 3, and so on to higher indices, one at a time. Since every rational number will occur in this sequence, it follows that the rational numbers are countable.

Virtually the same proof can be used to establish that the set of algebraic numbers is countable. But first we must know something about how many roots an algebraic equation can have. Recall that an algebraic number is one which satisfies some equation $f(x) = 0$ of the type

$$(1) \quad f(x) = a_n x^n + a_{n-1} x^{n-1} + a_{n-2} x^{n-2} + \cdots + a_2 x^2 + a_1 x + a_0 = 0,$$

with integer coefficients. We may presume that a_n is positive, for if it were negative we could multiply the equation by -1 without affecting its roots.

THEOREM C.1. *Any equation of the form* (1) *has at most n different roots.*

PROOF. Contrary to what is to be proved, let us suppose that eq. (1) has $n + 1$ different roots, say β_1, β_2, β_3, $\ldots$, β_n, β_{n+1}. We now use Theorem 7.2 (Chapter 7) or, rather, a slight variation on that result. The proof of that theorem assures us that $x - \beta$ is a factor of $f(x)$ if β is a root of $f(x) = 0$, whether or not β is a rational number. In case β is irrational, the quotient $q(x)$ has irrational coefficients, but that does not matter here. Thus in the present context, we see that $x - \beta_1$ is a factor of $f(x)$, say with quotient $q_1(x)$:

$$f(x) = (x - \beta_1)\, q_1(x).$$

Since β_2 is another root of $f(x) = 0$, we see that it must be a root of $q_1(x) = 0$, and so $x - \beta_2$ is a factor of $q_1(x)$, say with quotient $q_2(x)$:

$$q_1(x) = (x - \beta_2)\, q_2(x),$$
$$f(x) = (x - \beta_1)\, q_1(x) = (x - \beta_1)\, (x - \beta_2)\, q_2(x).$$

Continuing this process with β_3, β_4, $\ldots$, β_n, we observe that $f(x)$ can be factored into

$$(2) \qquad f(x) = (x - \beta_1)\, (x - \beta_2)\, (x - \beta_3) \cdots (x - \beta_n)\, q_n(x).$$

But $f(x)$ is of degree n, so $q_n(x)$ must be a constant; in fact, $q_n(x)$ must be a_n in order that this factoring shall agree with eq. (1).

Now consider the root β_{n+1}, which is different from all the other roots. From the fact that $f(\beta_{n+1}) = 0$, it follows by (2) that

$$(\beta_{n+1} - \beta_1)\, (\beta_{n+1} - \beta_2)\, (\beta_{n+1} - \beta_3) \cdots (\beta_{n+1} - \beta_n)\, a_n = 0,$$

which is impossible since the product of non-zero factors cannot be zero. Thus Theorem C.1 is proved.

THEOREM C.2. *The set of algebraic numbers is countable.*

PROOF. We say that the *index* of eq. (1) is

$$n + a_n + |a_{n-1}| + |a_{n-2}| + \cdots + |a_2| + |a_1| + |a_0|.$$

This is a positive integer since a_n is positive, and it is a straightforward generalization of the definition of index of a linear equation. Again, we may tabulate all equations for small values of the index, as shown in Table C 3.

TABLE C 3

Index	Equations
2	$x = 0$
3	$x^2 = 0, \quad 2x = 0, \quad \cdot x + 1 = 0, \quad x - 1 = 0$
4	$x^3 = 0, \, 2x^2 = 0, \, x^2 + x = 0, \, x^2 - x = 0, \, x^2 + 1 = 0,$
	$x^2 - 1 = 0, \quad 3x = 0, \quad 2x + 1 = 0,$
	$2x - 1 = 0, \quad x + 2 = 0, \quad x - 2 = 0$

As in the case of linear equations, we now list all new algebraic numbers arising from the equations of Table C 3. If we take them in order of magnitude for each index, we obtain the sequence

$$0; \; -1, 1; \; -2, -\frac{1}{2}, \frac{1}{2}, 2; \quad -3, \; -\frac{\sqrt{5}+1}{2}, \; -\sqrt{2}, \; -\frac{\sqrt{2}}{2},$$

(3)

$$-\frac{\sqrt{5}-1}{2}, \; -\frac{1}{3}, \frac{1}{3}, \frac{\sqrt{5}-1}{2}, \frac{\sqrt{2}}{2}, \; \sqrt{2}, \frac{\sqrt{5}+1}{2}, \; 3; \; -4, \ldots .$$

The number 0 comes from the one equation with index 2, the numbers -1 and $+1$ from the equations with index 3, the numbers -2, $-1/2$, $1/2$, 2 from the equations with index 4, and so on. The number of equations with any fixed index h is finite, because the degree n and the coefficients $a_n, \ldots, a_0$ are restricted to a finite set of integers. Also, by Theorem C.1 we know that each equation has at most n roots. Hence the sequence (3) will include all real algebraic numbers. It should be noted, however, that as we move to higher indices, although we can at each stage list all equations of any given index, we cannot continue to list the specific root forms as we have for the first few numbers in (3).

From Theorem C.2 we wish to draw the further conclusion that the set of real algebraic numbers between 0 and 1 is countable. This follows from a simple general principle, which we shall formulate as a theorem, about so-called subsets. A set M is called a subset of a set S if every element of M is an element of S.

THEOREM C.3. *Any infinite subset of a countable set is itself countable.*

PROOF. Let M be an infinite subset of a countable set S, say $S = \{a_1, a_2, a_3, a_4, \ldots\}$. Let a_{i_1} be the first element of S which is in M, a_{i_2} the second, and so on. Then M is the set:

$$M = \{a_{i_1}, a_{i_2}, a_{i_3}, \ldots\},$$

which clearly is countable.

Thus far, every infinite set we have considered has been countable. We now discuss a contrasting set which is uncountable.

THEOREM C.4. *The set of real numbers is uncountable.*

PROOF. In view of Theorem C.3, it will suffice to prove this for real numbers between 0 and 1; specifically for real numbers x satisfying $0 < x \leq 1$, so that 1 is included and 0 excluded. Suppose that the set of real numbers between 0 and 1 were countable, say

$$r_1, r_2, r_3, r_4, \ldots.$$

Write these numbers in decimal form, avoiding terminating decimals by using the infinite periodic form in all such cases (cf. Section 2.5). For example, the number 1/2 is to be written as $0.499999\cdots$, rather than 0.5. Thus we would have

$$r_1 = .a_{11}a_{12}a_{13}a_{14}a_{15}\cdots,$$
$$r_2 = .a_{21}a_{22}a_{23}a_{24}a_{25}\cdots,$$
$$r_3 = .a_{31}a_{32}a_{33}a_{34}a_{35}\cdots, \text{ etc.}$$

We now construct a number

$$\beta = .b_1b_2b_3b_4\cdots,$$

as follows. Let b_1 be any digit between 1 and 9, except that b_1 must be different from a_{11}. Similarly, let b_2 be any non-zero digit other than a_{22}. In general, let b_k be any non-zero digit other than a_{kk}. Hence the number β is different from r_1 (because they differ in the first decimal place), different from r_2 (because they differ in the second decimal place), and, in general, β is different from r_k (because they differ in the kth decimal place). Thus β is different from every one of the r's. But β is a real number between 0 and 1, and so we have a contradiction.

From this theorem, we can conclude that since the algebraic numbers between 0 and 1 are countable, but the real numbers between 0 and 1 are not, there must be real numbers which are not algebraic. These are the transcendental numbers, whose existence has thus been proved.

THEOREM C.5. *The set of real transcendental numbers is uncountable.*

PROOF. Suppose the real transcendental numbers were countable, say

$$t_1, t_2, t_3, t_4, \ldots .$$

Since by Theorem C.2 the real algebraic numbers are countable, say $a_1, a_2, a_3, a_4, \ldots$, the set of real numbers can be listed sequentially as

$$t_1, a_1, t_2, a_2, t_3, a_3, t_4, a_4, \ldots ,$$

contrary to Theorem C.4. Thus we have a contradiction and Theorem C.5 is established.

Finally, we note that Theorems C.2 and C.5 can be interpreted as saying that there are "more" transcendental numbers than algebraic numbers. The algebraic numbers can be listed in an infinite sequence, but there are too many transcendental numbers to allow such a sequential listing.

Problem Set 29

1. (a) List all linear equations with index 6, and (b) list all roots of these equations that are not roots of linear equations of lower index.

2. Prove that the set of all odd integers, positive and negative, is countable.

3. Prove that the set of polynomials $a + bx^4$, where a and b range over all natural numbers, is countable.

4. List all equations of index 5, and then verify the sequence (3) up to the element 3.

5. Prove that the set of numbers of the form $a + b\sqrt{3}$, where a and b range over all rational numbers, is countable.

6. Prove that, if a set A can be separated into two countable sets B and C, then A is countable.

7. Prove that the set of real numbers (strictly) between 0 and 0.1 is uncountable.

8. Prove that the set of all irrational numbers is uncountable.

Trigonometric Numbers

In Sections 5.1 and 5.2 certain numbers from trigonometry were shown to be irrational. By use of a more sophisticated method, we now prove the following general results.

THEOREM D.1. *Let θ be an angle whose measurement in degrees is a rational number. Suppose also that $0 < \theta < 90°$. Then $\cos \theta$, $\sin \theta$ and $\tan \theta$ are irrational numbers apart from the three exceptions:*

$$\cos 60° = \frac{1}{2}, \quad \sin 30° = \frac{1}{2}, \quad \tan 45° = 1.$$

A glance at the problem sets in Sections 5.1 and 5.2 shows that Theorem D.1 establishes all the examples of irrationality treated there, and many other cases not discussed in Chapter 5.

To prove this theorem, we begin by establishing two results from trigonometry. First we show that if θ is any angle and n a positive integer, then

(1) $2 \cos (n + 1)\theta = \{2 \cos n\theta\}\{2 \cos \theta\} - 2 \cos (n - 1)\theta.$

This identity can be obtained in the following way. Begin with the basic identities

$$\cos (A + B) = \cos A \cos B - \sin A \sin B,$$
$$\cos (A - B) = \cos A \cos B + \sin A \sin B,$$

and add these to obtain $\cos (A + B) + \cos (A - B) = 2 \cos A \cos B,$ or

$$\cos (A + B) = 2 \cos A \cos B - \cos (A - B).$$

129

Now replace A by $n\theta$ and B by θ, so that $A + B = (n+1)\theta$, $A - B = (n-1)\theta$, and the last relation becomes

$$\cos(n+1)\theta = 2\cos n\theta \cos\theta - \cos(n-1)\theta.$$

Multiplying this by 2 we obtain the desired identity (1).

Next we prove that for any positive integer n, $2\cos n\theta$ can be expressed as

(2) $2\cos n\theta = (2\cos\theta)^n + c_{n-1}(2\cos\theta)^{n-1} + \cdots + c_1(2\cos\theta) + c_0,$

where the coefficients c_{n-1}, c_{n-2}, $\cdots$, c_1, c_0 are integers. Before proving this, let us examine what it means for small positive integers n.

For $n = 1$, equation (2) has the simple form $2\cos\theta = 2\cos\theta$. For $n = 2$, the well-known identity $\cos 2\theta = 2\cos^2\theta - 1$ can be written as

$$2\cos 2\theta = (2\cos\theta)^2 - 2,$$

and this is another special form of equation (2). For $n = 3$, it is easy to prove by elementary trigonometry [or by use of identity (1) with $n = 2$] that

$$2\cos 3\theta = (2\cos\theta)^3 - 3(2\cos\theta).$$

This is again of the form of equation (2), with $n = 3$, $c_2 = 0$, $c_1 = -3$ and $c_0 = 0$.

Of course it must be understood that the values of the constants c_0, c_1, etc. in equation (2) are different for different values of n. For example, using identity (1) with $n = 3$ we get the next case

$$2\cos 4\theta = (2\cos\theta)^4 - 4(2\cos\theta)^2 - 2.$$

Thus for $n = 4$ the constants in equation (2) have the values $c_3 = 0$, $c_2 = -4$, $c_1 = 0$ and $c_0 = -2$.

Having discussed how equation (2) is to be interpreted, we now prove it in general by mathematical induction. This involves showing that, if there is an expression like (2) for $2\cos n\theta$, then there is a similar kind of expression for $2\cos(n+1)\theta$. In addition to presuming a formulation (2) for $2\cos n\theta$, we shall presume also a formulation at the preceding stage,

(3) $2\cos(n-1)\theta = (2\cos\theta)^{n-1} + b_{n-2}(2\cos\theta)^{n-2} + \cdots$
$$+ b_1(2\cos\theta) + b_0,$$

where the coefficients are integers. Now if equations (2) and (3) are substituted into the identity (1), we get

$$2 \cos (n + 1)\theta$$
$$= 2 \cos \theta[(2 \cos \theta)^n + c_{n-1}(2 \cos \theta)^{n-1} + \cdots + c_1(2 \cos \theta) + c_0]$$
$$\quad - [(2 \cos \theta)^{n-1} + b_{n-2}(2 \cos \theta)^{n-2} + \cdots + b_1(2 \cos \theta) + b_0]$$
$$= (2 \cos \theta)^{n+1} + c_{n-1}(2 \cos \theta)^n + (c_{n-2} - 1)(2 \cos \theta)^{n-1}$$
$$\quad + (c_{n-3} - b_{n-2})(2 \cos \theta)^{n-2} + \cdots + (c_0 - b_1)(2 \cos \theta) - b_0.$$

Since all the coefficients are integers, this is again of the form (2), but now for $2 \cos (n + 1)\theta$. Hence, by mathematical induction, there is an equation of the form (2) for all values of n.

Now we are in a position to prove Theorem D.1. If θ is an angle whose degree measure is a rational number, then there is an integer n such that $n\theta$ is an integer multiple of $360°$. For example, if θ is $23/7$ degrees, we could take $n = 7 \cdot 360$. In general, if θ equals a/b degrees, where a and b are integers, we take $n = 360b$ so that $n\theta = 360a$. Then $\cos n\theta = 1$ because $n\theta$ is a multiple of $360°$. Substituting this into equation (2) we obtain, after rearranging,

$$(4) \quad (2 \cos \theta)^n + c_{n-1}(2 \cos \theta)^{n-1} + \cdots + c_1(2 \cos \theta) + c_0 - 2 = 0.$$

But this means that for the angle θ under consideration, $2 \cos \theta$ is a root of the polynomial equation with integer coefficients

$$x^n + c_{n-1}x^{n-1} + c_{n-2}x^{n-2} + \cdots + c_1 x + c_0 - 2 = 0.$$

Now by Corollary 1 on page 60, any rational root of this equation is an integer. So if $2 \cos \theta$ is rational, it is an integer. However, $\cos \theta$ is at most 1 and at least -1, hence $2 \cos \theta$ is at most 2 and at least -2. In Theorem D.1 it is assumed that $0 < \theta < 90°$, so $\cos \theta$ is between 1 and 0. Thus $2 \cos \theta$ is between 2 and 0. The only integer between 2 and 0 is 1, so if $2 \cos \theta$ is rational, we conclude that

$$2 \cos \theta = 1, \quad \cos \theta = \tfrac{1}{2}, \quad \theta = 60°.$$

This argument establishes Theorem D.1 for $\cos \theta$.

As to $\sin \theta$, if θ is rational in degrees and $0 < \theta < 90°$, then its complement $90° - \theta$ is also rational in degrees, and furthermore, $0 < 90° - \theta < 90°$. By elementary trigonometry we know that

$$\sin \theta = \cos (90° - \theta)$$

and hence if $\sin\theta$ is rational, so is $\cos(90° - \theta)$. But from the part of Theorem D.1 just established, $\cos(90° - \theta)$ is rational only if $90° - \theta = 60°$, i. e. only if $\theta = 30°$. Thus we have proved Theorem D.1 for $\sin\theta$.

Turning now to $\tan\theta$, we first make this observation: *If the degree measure of α is rational, and if* $0 < \alpha < 180°$, *then* $\cos\alpha$ *is rational only in the three cases* $\alpha = 60°$, $\alpha = 90°$, $\alpha = 120°$. This simple extension of Theorem D.1 for the cosine of angles between 0 and 180 degrees follows from the facts that $\cos 90° = 0$ and $\cos\alpha = -\cos(180° - \alpha)$.

Now we use the identity

$$(5) \qquad \cos 2\theta = \cos^2\theta - \sin^2\theta = \frac{\cos^2\theta - \sin^2\theta}{\cos^2\theta + \sin^2\theta} = \frac{1 - \tan^2\theta}{1 + \tan^2\theta}.$$

Assume that $\tan\theta$ is rational; then $\tan^2\theta$ is rational, the last fraction in (5) is rational, and so $\cos 2\theta$ is rational. Since θ satisfies $0 < \theta < 90°$, 2θ satisfies $0 < 2\theta < 180°$. From the observation of the preceding paragraph it follows that the only possibilities for 2θ are $2\theta = 60°$, $2\theta = 90°$, $2\theta = 120°$; i.e., $\theta = 30°$, $\theta = 45°$, $\theta = 60°$. But

$$\tan 30° = \sqrt{3}/3 \quad \text{and} \quad \tan 60° = \sqrt{3}$$

are irrational. On the other hand $\tan 45° = 1$, and this is rational. Thus the proof of Theorem D.1 is complete.

Answers and Suggestions
to Selected Problems

Set 1

1. (a) False: $1 + 1 = 2$.
   (b) True.
   (c) False: $1 - (-1) = 2$.
   (d) True.
   *(e) False: $2^1 + 2^2 = 6$, and 6 is not an integral power of 2.
2. Eight, namely, 1, 2, 3, 5, 6, 10, 15, 30.
3. Five, namely 1, 2, 4, 8, 16.
4. 4.
5. 53, 59, 61, 67, 71, 73, 79, 83, 89, 97.
*6. *Suggestion.* Find a good notation for numbers which are exact multiples of a given number d.

Set 2

1. Yes; $q = -7$.
2. Yes; $q = -7$.
3. Yes; $q = 7$.
4. No.
5. Yes; $q = -35$.
6. Yes; $q = 0$.
7. No.
8. Yes; $q = 1$.
9. No, because q is not unique.
10. Yes.
11. Yes.

Set 3

1. (a), (b), and (f) are true; (c), (d), and (e) are false.
2. True in all cases.
3. (a), (c), and (d) are true; (b) and (e) are false.
4. (a), (b), (c), and (d) are true; (e) is false.

133

Set 4

6. (a) not closed, (b) closed, (c) closed, (d) not closed, (e) closed, (f) closed, (g) closed.

Set 6

1. (a) 0.25;   (b) 0.015;   (c) 0.8025;
   (d) 0.0112;   (e) 2.816;   (f) 1.2596.

Set 7

2. (a) False, for example in case $b = 10$;
   (b) True;
   (c) False, for example in case $b = 10$;
   (d) False, for example in case $b = 7$;
   (e) False, for example in case $b = 7$;
   (f) True.
3. (a) False, for example in the case of the fraction 3/6;
   (b) True;
   (c) False, for example in the case of the fraction 3/6.
4. If $ab = 0$, then $a = 0$ or $b = 0$.
5. (b) Yes.

Set 8

1. (a) 1/9;   (d) 9978/9990 = 1663/1665;
   (b) 17/3;   (e) 1/9900;
   (c) 3706/9900 = 1853/4950;   (f) 1.

Set 9

1. (a) 0.12;   (b) 0.3;   (c) 4.8;   (d) 10.0.
2. (a) 0.72999 $\cdots$;   (b) 0.0098999 $\cdots$;   (c) 12.999 $\cdots$.
3. Rational numbers a/b (in lowest terms) with the property that b is divisible by no prime other than 2 and 5 and with $a \neq 0$.
4. None.

Set 10

7. Rational.

Set 11

1. $\sqrt{3}$ and $-\sqrt{2}$ will do.
2. $\sqrt{2}$ and $\sqrt{2}$ will do.
3. $\sqrt{2}$ and $\sqrt{3}$ will do.
4. $\sqrt{2}$ and $\sqrt{2}$ will do.
5. $\sqrt{3}$ and $1/\sqrt{2}$ will do.

Set 12

1. (a) $n = 3$, $c_3 = 15$, $c_2 = -23$, $c_1 = 9$, $c_0 = -1$;
   (b) $n = 3$, $c_3 = 3$, $c_2 = 2$, $c_1 = -3$, $c_0 = -2$;
   (c) $n = 3$, $c_3 = 2$, $c_2 = 7$, $c_1 = -3$, $c_0 = -18$;
   (d) $n = 4$, $c_4 = 2$, $c_3 = 0$, $c_2 = -1$, $c_1 = -3$, $c_0 = 5$;
   (e) $n = 5$, $c_5 = 3$, $c_4 = 0$, $c_3 = -5$, $c_2 = 6$, $c_1 = -12$, $c_0 = 8$;
   (f) $n = 4$, $c_4 = 1$, $c_3 = 0$, $c_2 = -3$, $c_1 = -5$, $c_0 = 9$.
2. (a) yes; (b) yes; (c) yes; (d) no; (e) yes; (f) no.
4. *Suggestion.* Multiply the equation by the product $b_3b_2b_1b_0$.

Set 13

2. *Suggestion.* Use Theorem 4.1 and one of the results of Problem 1.
7. *Suggestion.* 2/2 is a root of $x^2 - 1 = 0$, for example.

Set 15

1. (a) *Suggestion.* Replace θ by $40°$ in eq. (5), and use the fact that $\cos 120° = -1/2$.
   (b) *Suggestion.* Use the result of Problem 1(a) and eq. (8).
   (c) *Suggestion.* Use eq. (8), part 1, with $\theta = 10°$.
   (d) *Suggestion.* Use the result of Problem 1(a) and the identity $\cos\theta = \sin (90° - \theta)$.
3. (a) *Suggestion.* Replace A by 3θ and B by 2θ in eq. (1), and use eqs. (3), (4), (5), and (7).
4. (a), (b), (c), (d), (i), (k) are rational.

Set 16

1. (a) *Suggestion.* Use $\cos 30° = \sqrt{3}/2$.
   (c) *Suggestion.* Use $\cos 45° = \sqrt{2}/2$.
   (d) *Suggestion.* Use the facts that $\cos 40°$ is irrational, and that $\cos 2 \cdot 35° = \cos 70° = \cos (90° - 20°) = \sin 20°$, etc.
3. (b) Yes.

Set 17

3. *Suggestion*. Recall that $\log m + \log n = \log mn$.

4. *Suggestion*. Make use, among other things, of Example 3 in the text.

Set 18

1. (a) *Suggestion*. It is a root of $x^2 - 3 = 0$.

   (b) *Suggestion*. It is a root of $x^3 - 5 = 0$.

   (c) *Suggestion*. It is a root of $x^4 - 10x^2 + 1 = 0$. See eq. (5) of Chapter 4.

   (d) *Suggestion*. See eq. (6) of Chapter 5.

Set 21

5. (a) False, for example if $r = -2$ and $s = -3$;

   (b) False, for example if $r = 4$, $s = 3$, and $c = -2$;

   (c) True;

   (d) True;

   (e) True;

   (f) False, for example if $\lambda = 2/5$;

   (g) True.

6. $-10 < \lambda < 10$.

8. (b) Yes. The difference is that $u - v$ can be 0 in (b) but not in (a).

Set 22

2. (a) 1, (b) 3, (c) 4, (d) 6, (e) 5, (f) 7, (g) 3, (h) 31, (i) -2, (j) -22.

Set 23

1. 2/1, 3/2, 5/3, 7/4, 9/5, 10/6, 12/7, 14/8, 16/9, 17/10.

2. 3/1, 6/2, 9/3, 13/4, 16/5, 19/6, 22/7, 25/8, 28/9, 31/10.

3. *Suggestion*. Deduce this from Theorem 6.3.

*5. *Suggestion*. Consider the case where $\lambda = \sqrt{2}$ and $n = 4$, and establish that there is no fraction $m/4$ in lowest terms (i.e., with m odd), such that

$$-\frac{1}{8} < \lambda - \frac{m}{4} < \frac{1}{8}.$$

Set 24

1. $n = 4$, $m = 7$.

2.

	(a)	(b)	(c)	(d)	(e)	(f)	(g)	(h)	(i)	(j)	(k)	(l)	(m)	(n)
n	2	3	4	4	1	3	5	5	5	5	1	1	1	7
m	3	5	7	7	1	4	7	7	7	7	3	3	3	22

Set 25

1. *Suggestion.* Take the integer just larger than λ and the integer just smaller.
2. *Suggestion.* Show that the exception is m/n with $n = 1$, where m is, of the two integers just less than and just greater than λ, that one which is farther away.
3. (a) 3/2 and 4/3 will do.
   (b) 3/2 and 5/3 will do.
   (c) 7/3 and 9/4 will do.
4. (a) All of them.
   (b) 1/1, and also 14/10 provided it is taken in the reduced form 7/5.
5. (a) 3/1, 31/10, 314/100; (b) 3/1.
*6. *Suggestion.* Prove that the inequalities of Theorem 6.5 are false for $\lambda = 3/5$ and any m/n with $n > 5$, as follows: $\lambda - m/n$ is either positive or negative. If positive, show it is at least $1/5n$; if negative, at most $-1/5n$.
*7. (a) *Suggestion.* Use the Fundamental Theorem of Arithmetic, as given in Appendix B, to prove that the given rational numbers are unequal.
   (b) *Suggestion.* Prove that the inequalities of Theorem 6.5 cannot hold for any rational number m/n with n larger than b.

Set 26

1. (b) There are none.
2. (b) 1/1, 2/1, 3/2.
3. (b) 1/1, 2/1.

Set 27

1. 2, 2, 8, and 10^{-1}.
2. No.
4. (b) $|x - 7| = x - 7$ if $x \geqq 7$; $|x - 7| = -x + 7$ if $x \leqq 7$.
5. (a) $x = -1$; (b) $x = 2$; (c) $x = 7$ and $x = -7$; (d) all values of x.

Set 28

2. $\alpha^7 - \beta^7 = (\alpha - \beta)(\alpha^6 + \alpha^5\beta + \alpha^4\beta^2 + \alpha^3\beta^3 + \alpha^2\beta^4 + \alpha\beta^5 + \beta^6)$.
3. *Suggestion.* Any root of $f(x) = 0$ is also a root of $f(x)\,g(x) = 0$.

Set 29

1. (a) $5x = 0, 4x \pm 1 = 0, 3x \pm 2 = 0, 2x \pm 3 = 0, x \pm 4 = 0$;
   (b) $-4, -3/2, -2/3, -1/4, 1/4, 2/3, 3/2, 4$.
2. For example, $1, -1, 3, -3, 5, -5, 7, -7, 9, -9, \ldots$.
3. *Suggestion.* Define the *index* of $a + bx^4$ as $a + b$; then observe that there is only a finite number of polynomials of any given index, and enumerate them all.

4. $x^4 = 0$, $2x^3 = 0$, $x^3 \pm 1 = 0$, $x^3 \pm x = 0$, $x^3 \pm x^2 = 0$, $3x^2 = 0$, $2x^2 \pm 1 = 0$, $x^2 \pm 2 = 0$, $2x^2 \pm x = 0$, $x^2 \pm x \pm 1 = 0$, $x^2 \pm 2x = 0$, $4x = 0$, $3x \pm 1 = 0$, $2x \pm 2 = 0$, $x \pm 3 = 0$.

5. *Suggestion.* All these numbers are algebraic; use Theorem C.3.

6. *Suggestion.* Let b_1, b_2, b_3, ... be a sequential listing of the elements of B, and let c_1, c_2, c_3, ... be such a listing of the elements of C; then A can be written sequentially as

$$b_1, c_1, b_2, c_2, b_3, c_3, \ldots .$$

7. *Suggestion.* Follow the proof of Theorem C.4; but the numbers a_{11}, a_{21}, a_{31}, ... are all zero in the present context. Construct the desired not-listed number by choosing $b_1 = 0$, $b_2 \neq a_{12}$ and $b_2 \neq 0$, $b_3 \neq a_{23}$ and $b_3 \neq 0$, and, in general, $b_i \neq a_{i-1,i}$ and $b_i \neq 0$

INDEX